P9-BYH-390

Merriam-Webster's Guide to Everyday Math

A Home and Business Reference

Brian Burrell

Illustrations by the author

Merriam-Webster, Incorporated
Springfield, Massachusetts

A GENUINE MERRIAM-WEBSTER

The name *Webster* alone is no guarantee of excellence. It is used by a number of publishers and may serve mainly to mislead an unwary buyer.

Merriam-Webster™ is the name you should look for when you consider the purchase of dictionaries and other fine reference books. It carries the reputation of a company that has been publishing since 1831 and is your assurance of quality and authority.

Copyright © 1998 by Brian Burrell

Library of Congress Cataloging-in-Publication Data

Burrell, Brian, 1955–
    Merriam-Webster's guide to everyday math : a home and business reference / Brian Burrell.
        p.   cm.
    Includes index.
    ISBN 0-87779-621-1
    1. Mathematics—Handbooks, manuals, etc.   I. Title.
QA40.B868   1998
510—dc21   98-18147
                                                            CIP

Printed and bound in the United States of America

2345RRD0201009998

Contents

Preface

Merriam-Webster's Guide to Everyday Math is a handy reference to the math concepts you are likely to encounter at home and at work. It is intended for math teachers, math students, and especially for *former* math students—in other words, for everyone. It contains detailed explanations of math-related ideas that appear in the financial pages of newspapers, in popular magazines, and in the nightly news. If you have ever found yourself rummaging around for an old math textbook in search of explanations of concepts you once understood (or never grasped very well), you don't have to look any further. It's all here.

Because mathematics is the common language of the sciences, it is only natural that topics from physics, chemistry, meteorology, geography, optics, and engineering should appear in these pages, along with business applications. Scientific terms and concepts such as dew point, Richter scale, light-year, longitude, and decibel share equal billing with money-related concepts such as dollar-cost averaging, compound interest, basis points, and gambling odds.

Some of these concepts are, of course, more complex than others. The topics covered here range in difficulty from the very basic—those which use nothing more than addition and subtraction—to demanding calculations that require such tools as logarithms, exponents, and trigonometric functions. Fortunately, most of these can be broken down into simple steps, which is how they are presented here.

I have tried to explain every concept clearly and even conversationally. Topics are explained in series of numbered paragraphs, each of which covers a single concept or definition. Hundreds of examples are provided; these are indented and set off with a diamond-shaped bullet (♦). Sidebars containing interesting applications and anecdotes are interspersed throughout the book. When first encountered and defined, mathematical terms appear in italics (e.g., *percentile rank*). A list of practical applications (e.g., board feet, pH scale, braking distance) is printed on the inside front and back covers, providing a concise directory to the entire book.

It is only natural that anyone who has not taken a math course recently may need to refresh his or her memory on certain points. For this reason, the first two chapters contain a review of high school math. Chapter 1 explains the rules of arithmetic, including decimals, percents, and fractions as well as logarithms, exponents, and roots, while Chapter 2 covers the basic concepts of algebra, geometry, and trigonometry. Many of the computations worked out in the examples require a calculator of some sort; consequently I have provided some tips and suggestions for using calculators at the end of Chapter 1, and here and there in later chapters.

What most readers will think of as "everyday math" comes next: Chapter 3 covers household math; Chapter 4 concerns the universe, the world, and the environment; Chapter 5 explains the laws of chance; and Chapter 6 deals with the fundamentals of money and personal finance. The Appendix contains a comprehensive set of conversion tables and geometric formulas, a table of mathematical symbols, and some useful rules of thumb.

This volume is a revised and expanded version of *Merriam-Webster's Pocket Guide to Business and Everyday Math*. New, or greatly expanded, sections include those on geography, astronomical distances, measuring time, measuring sound, and, in a concession to current trends, casino games and betting strategies. Although it contains much that is new, it still owes a great debt to the many individuals who are cited in the *Pocket Guide* for their helpful suggestions and comments. My thanks go out to them again. At Merriam-Webster, Mark A. Stevens edited the manuscript and oversaw the production process, Michael D. Roundy painstakingly checked the entire text for accuracy, Jocelyn White Franklin and Michael Shally-Jensen provided production assistance, Daniel J. Hopkins proofread the typeset pages, and Robie Grant prepared the index.

Brian Burrell

Merriam-Webster's
Guide to
Everyday Math

1 Numbers and Calculating

It would be impossible to talk about size, quantity, amount, distance, or duration without using numbers.

Numbers should be distinguished from digits. The ten digits, or Hindu-Arabic numerals—0, 1, 2, 3, 4, 5, 6, 7, 8, 9—are the symbols used to represent numbers. Thus a digit is a symbol, whereas a number is an expression of quantity and order.

There are many types of numbers, and as many names with which to describe them. This section covers the most common classifications and gives examples to show how they come up in everyday settings.

Number Systems

1. The numbers that a child first learns are the *counting numbers* or *natural numbers:* 1, 2, 3, 4, 5, . . .

2. The *whole numbers* or *cardinal numbers* consist of the counting numbers and zero: 0, 1, 2, 3, 4, 5, . . .

3. The set of *integers* refers to the whole numbers and their additive inverses (the negative counting numbers). This set can be represented as

$$\{. . . -3, -2, -1, 0, 1, 2, 3, . . .\}$$

It is often used to label a simple number line.

4. A *fraction* is a *quotient* of two integers—the result of dividing one integer by another. Fractions are often referred to as *rational numbers* because they are *ratios* of integers. The integer on top is called the *numerator;* the one on the bottom is the *denominator.*

- In the fraction ¾, 3 is the numerator and 4 is the denominator. The implied meaning is three parts out of a whole that has been divided into four parts. It can also be thought of as a quotient: 3 divided by 4.

5. Fractions and integers do not account for every number on the number line. Between any two fractions there can always be found an *irrational number,* a number that cannot be represented as a fraction. The three most commonly encountered irrational numbers are π(*pi*), the ratio of the circumference of a circle to its diameter; the square root of 2; and the number *e,* which occurs naturally in calculations involving compound interest (see Chapter 6).

6. The types of numbers discussed above—integers, fractions, and irrational numbers—account for every possible value or place on the standard number line, or real number line. That is, the *real numbers* consist of the counting numbers, the negative counting numbers, zero, all fractions, and numbers such as $\sqrt{2}$, $\sqrt{3}$, *e,* and π that correspond to real places on the number line but cannot be expressed as fractions.

Fractions, Decimals, Percents

In addition to the many ways of *classifying* a number, there are also many ways to *represent* a number. The most familiar are fractions, decimals, and percents. Each of these has its advantages and limitations.

FRACTIONS

1. There are three types of fractions:

  i. *simple fractions* (such as ½ or ¾), in which the numerator is less than the denominator,
  ii. *improper fractions* (such as ¹¹⁄₃ or ⁴³⁄₅), in which the numerator is equal to or greater than the denominator, and
  iii. *mixed numbers* (such as 2½), in which an improper fraction is expressed as a whole number with a simple fractional remainder.

An improper fraction can be converted to a mixed number by dividing the denominator into the numerator and leaving the remainder in fraction form.

♦ $^{23}/_4$ is equivalent to $5^3/_4$ because 23 divided by 4 is 5 with 3 left over. The part "left over" is called the *remainder*, and it is expressed as a fraction using the original denominator.

A mixed number can be converted to an improper fraction by multiplying the whole number by the denominator, and then adding the numerator to generate a new numerator. The denominator remains the same.

$$\bullet\ 5\frac{2}{3} = \frac{(3 \times 5) + 2}{3} = \frac{17}{3}$$

2. *Equivalent fractions* have the same numerical value (and thus the same decimal representation), but express equal portions in terms of different base amounts. One-half (as in a half-gallon of milk) is equivalent to two-fourths (or two quarts)—the same amount of milk, but different packaging. Of course the packaging may be relevant. Two-and-a-half dollars (which is $^5/_2$) is equivalent to ten quarters (or $^{10}/_4$), which would be of more use at a pay phone.

3. The numerator and denominator of any proper or improper fraction may be multiplied by the same number in order to produce an equivalent fraction.

♦ $^4/_5$ becomes $^8/_{10}$, $^{12}/_{15}$, and $^{16}/_{20}$ when both its numerator and denominator are doubled, tripled, and quadrupled, respectively.

The reverse of this process is called *reducing* the fraction.

♦ $^{28}/_{84}$ can be reduced to $^{14}/_{42}$ by dividing both numerator and denominator by 2. $^{14}/_{42}$ can be reduced to $^7/_{21}$ by the same process. Finally, $^7/_{21}$ can be reduced to $^1/_3$ by dividing numerator and denominator by 7. Because no more reducing is possible, the result is said to be in *reduced form*.

4. Fractions may be used to express proportional amounts or *ratios* (see Chapter 2). Thus fractions are used both to establish increments of measurement and to establish comparisons of size or quantity. They are frequently encountered in measurements, because in the real world (unlike on blackboards) nothing is exact.

5. Comparing Fractions
If two fractions have:

i. *the same denominator,* then the one with the larger numerator is larger.

◆ $\dfrac{7}{8}$ is greater than $\dfrac{5}{8}$; $\dfrac{21}{3}$ (or 7) is less than $\dfrac{23}{3}$.

ii. *the same numerator,* then the one with the smaller denominator is larger.

◆ $\dfrac{5}{4}$ is less than $\dfrac{5}{3}$; $\dfrac{7}{8}$ is greater than $\dfrac{7}{13}$.

iii. *different numerators and denominators,* then convert them to equivalent fractions with the same denominator, and compare numerators.

◆ $\dfrac{7}{8}$ is less than $\dfrac{8}{9}$ because $\dfrac{7}{8} = \dfrac{7 \times 9}{8 \times 9} = \dfrac{63}{72}$, and $\dfrac{8}{9} = \dfrac{8 \times 8}{9 \times 8}$ $= \dfrac{64}{72}$.

The relative size of two fractions can also be easily compared by converting them to decimal form.

DECIMALS
1. In decimal representation, a number consists of a string of one or more digits that occupy *place values.*

- ◆ The number 735 means 7 *hundreds* plus 3 *tens* plus 5 *ones.*
- ◆ The number 0.49 means 4 *tenths* plus 9 *hundredths.*

The decimal point fixes the place value of each digit: place values to the left of the decimal point indicate increasing powers of 10; place values to the right indicate fractional powers of 10 (tenths, hundredths, thousandths, etc.). In the case of a whole number, the decimal point, which would follow the last (or "ones" place) digit, is omitted. Thus the number 256 represents 2 hundreds plus 5 tens plus 6 ones. The number 0.075 represents 7 hundredths plus 5 thousandths. Every real number has a decimal representation. The simplest real numbers—the whole-valued numbers such as $-3, -2, -1, 0, 1, 2, 3$, etc.—are already in decimal form.

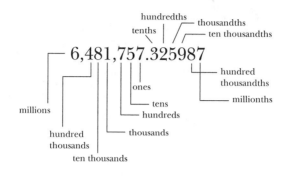

Φ ———————————————————————————— Φ

NAMES FOR NUMBERS

When writing a decimal number in words, the word *and* takes the place of the decimal point. It should not be inserted anywhere else. Thus 256 is "two hundred fifty-six," and not "two hundred and fifty-six." The reason becomes apparent when fractions are involved. "Two hundred and fifty-six thousandths" means 200.056, not 0.256. Obviously, the difference a misplaced *and* can make is crucial. It might help to think about how you write out an amount on a personal check. $124.52 is "one hundred twenty-four and $^{52}/_{100}$ dollars"; the *and* separates the dollars from the cents.

Φ ———————————————————————————— Φ

2. Converting a Fraction to a Decimal To express a fraction as a decimal, divide the denominator into the numerator. (Notice how the division symbol ÷ itself resembles a fraction.) Thus ½ can be written as ½ or 1 ÷ 2. The decimal form is 0.5.

3. Converting a Decimal to a Fraction Decimals are converted to fractions by writing the appropriate power of 10 as a denominator. Decimals in which all digits (other than zero) fall to the right of the decimal point convert to proper fractions. Those with digits (other than zero) to either side of the decimal point convert to improper fractions or mixed numbers.

♦ 0.15 converts to $\dfrac{15}{100}$.

♦ 2.15 converts to $\dfrac{215}{100}$ or $2\dfrac{15}{100}$.

It is useful to note that the term *decimal* is short for *decimal fraction,* which designates a fraction whose denominator is a power of 10 (see pages 41–43 for a discussion of powers of 10). Thus 0.5, 0.35, and 0.125 are all decimal fractions whose fractional equivalents are 5/10, 35/100, and 125/1000, respectively.

4. A fraction may convert to one of two types of decimals—a *terminating decimal* or a *repeating decimal.*

♦ 7/8, or 7 ÷ 8, is equivalent to 0.875, a decimal which terminates.

♦ 5/6, or 5 ÷ 6, is equivalent to the decimal number 0.833333 . . . , which repeats.

The ellipsis (. . .) indicates the repetition of the digits. Another way to express repetition is to use a horizontal bar over the digits that repeat.

♦ $\dfrac{1}{3}$ can be written as $0.\overline{3}$.

♦ $\dfrac{5}{12}$ can be written as 0.416666 . . . , or as $0.41\overline{6}$.

♦ $\dfrac{4}{33}$ can be written as 0.1212121212 . . . , or as $0.\overline{12}$.

5. Algebra lets us convert repeating decimals to fractions. The key to the method is to choose a multiplier based upon the number of digits that repeat. If one digit repeats, the multiplier is 10; if two repeat, it is 100; if three repeat, 1000; if four, 10,000; and so on.
    The method consists of four steps:

   i. Let x equal the repeating decimal.
  ii. Multiply both sides of the equation by the multiplier.
 iii. Subtract equation 1 from equation 2 to eliminate the repeating digits.
 iv. Divide each side of the resulting equation by the number that multiplies x.

- Convert $0.\overline{3}$ to a fraction.
    1. Let $x = 0.3333\ldots$
    2. Then $10x = 3.3333\ldots$
    3. Thus $10x - x = (3.3333\ldots) - (0.3333\ldots)$ or $9x = 3$.
    4. $9x = 3$ when divided by 9 becomes $x = \dfrac{1}{3}$.

Table 1.1. Common Repeating Decimals and Fractional Equivalents

$0.3333\ldots = \dfrac{1}{3}$	$0.5555\ldots = \dfrac{5}{9}$
$0.6666\ldots = \dfrac{2}{3}$	$0.7777\ldots = \dfrac{7}{9}$
$0.1666\ldots = \dfrac{1}{6}$	$0.8888\ldots = \dfrac{8}{9}$
$0.8333\ldots = \dfrac{5}{6}$	$0.0909\ldots = \dfrac{1}{11}$
$0.1111\ldots = \dfrac{1}{9}$	$0.1818\ldots = \dfrac{2}{11}$
$0.2222\ldots = \dfrac{2}{9}$	$0.08333\ldots = \dfrac{1}{12}$
$0.4444\ldots = \dfrac{4}{9}$	$0.41666\ldots = \dfrac{5}{12}$

6. If some initial digits do not repeat, start with an extra step: Multiply by a power of 10 that shifts the nonrepeating digits to the left of the decimal point.

- If $x = 0.416666\ldots$, then
    1. $100x = 41.6666\ldots$
    2. $10 \cdot 100x = 1000x = 416.6666\ldots$
    3. $1000x - 100x = (416.6666\ldots) - (41.6666\ldots)$
       $= 416 - 41 = 375$.
    4. $900x = 375$, or $x = \dfrac{375}{900}$, which reduces to $\dfrac{5}{12}$.

7. A quicker way to convert a repeating decimal to a fraction is to write a fraction in which the numerator consists of the digits that repeat, and the denominator consists of the same number of 9's.

- $0.12121212\ldots$ converts to $\dfrac{12}{99}$, which reduces to $\dfrac{4}{33}$.

This method only works when the pattern of repetition begins immediately after the decimal point.

8. Irrational numbers also have decimal forms, although they turn out to be *nonterminating* and *nonrepeating*. This means that a number such as $\sqrt{2}$ has a decimal form consisting of a string of digits that never repeats a pattern. For practical purposes such decimals are always *rounded off* (see page 20). A scientific calculator, for example, may express $\sqrt{2}$ as 1.41421356, rounding to the last digit because it can only display nine characters. For most purposes, however, nine decimal places are more than adequate.

PERCENTS AND PERCENTAGES

The word *percent* means "out of one hundred." The percent symbol, %, like the division symbol, ÷, represents a quotient. (Notice how the two symbols even look alike.) 70% means 70 out of 100, or $^{70}\!/_{100}$, or 0.7. Thus a percent is a means of describing a portion of a whole that has been divided into 100 equal parts.

The amount to which a percent is applied is referred to as the *base*, and the number derived from a percent and a base is called a *percentage*. In order to find a percentage, multiply the percent (in decimal form) times the base amount.

◆ *25 percent of 120* is written as 25% × 120, or as $^{25}\!/_{100}$ × 120, which is 30. In this instance, 30 is a percentage of 120: $^{30}\!/_{120} = 25\%$.

1. Converting a Percent to a Decimal A percent can be converted to a decimal by moving the decimal point two places to the left and removing the percent sign. In the expression 5%, the implied position of the decimal point is after the 5—that is, 5 is the same as 5.0—and thus 5% is equivalent to 0.05. Similarly, 86% is equivalent to 0.86.

To convert a percent that contains a mixed number (a whole number and a fraction) to a decimal, the mixed number must first be converted to its decimal form. Then the decimal point can be moved two places to the left.

◆ $84\frac{1}{2}\%$ is equivalent to 84.5%, or 0.845.

◆ An interest rate of 6¾% is equivalent to 6.75%, which converts to 0.0675.

2. Converting a Decimal to a Percent The rule for converting any number in decimal form to percent form is to move the decimal point two places to the right and write a percent sign.

◆ 0.35 is the same as 35%.

◆ 1.6 is 160%.

◆ 0.015 is 1.5%, or $1\frac{1}{2}$%.

3. Converting a Repeating Decimal to a Percent Like an ordinary decimal, a repeating decimal can be converted to a percent by moving the decimal point two places to the right. But this still leaves a repeating decimal, which should not be followed by a percent sign. The repeating part of the decimal can be rounded or shortened (the hundredths' place is adequate for most purposes), or expressed as a fraction. The result will be a *mixed-number percent*.

◆ 0.666 . . . is equivalent to ⅔. Moving the decimal point two places to the right leaves 66.666 . . . , or 66⅔. Thus $0.\overline{6}$ is approximately equal to 66.67 percent and exactly equal to 66⅔%.

4. Converting a Fraction to a Percent Fractions should be converted to decimal form before being converted to percent form. Once the decimal point has been moved two places to the right, the decimal can be converted to a fractional percent. See Table 1.2.

◆ $1\frac{3}{8}$ equals 1.375, which converts to 137.5%, or $137\frac{1}{2}$%.

◆ Mortgage interest rates are listed in increments of ⅛ of a percent. 7⅜%, for example, is equivalent to 7.325%, which in decimal form is 0.07325.

5. If a fraction takes the form of a repeating decimal, the repeating portion of the decimal can be converted back to fraction form after the decimal point has been moved.

◆ ⅓ equals $0.33\overline{3}$, which rounds to 33.3%, or can be expressed exactly as 33⅓%.

6. Converting a Percent to a Fraction Because *percent* means "out of 100," any percent can be converted to a fraction by placing it over 100.

With a mixed-number percent, however, the best way to proceed is first to convert to a decimal, and then to a fraction.

- $48\% = \dfrac{48}{100}$

- $8\dfrac{1}{2}\% = 8.5\% = 0.085 = \dfrac{85}{1000}$

Table 1.2. Common Fractions with Decimal and Percent Equivalents

fraction	decimal	percent
½	0.5	50%
¼	0.25	25%
¾	0.75	75%
⅓	$0.33\overline{3}$	33⅓%
⅔	$0.66\overline{6}$	66⅔%
⅕	0.2	20%
⅖	0.4	40%
⅗	0.6	60%
⅘	0.8	80%
⅙	$0.16\overline{6}$	16⅔%
⅚	$0.83\overline{3}$	83⅓%
⅛	0.125	12½%
⅜	0.375	37½%
⅝	0.625	62½%
⅞	0.875	87½%
⅑	$0.11\overline{1}$	11⅑
2/9	$0.22\overline{2}$	22²⁄₉%
4/9	$0.44\overline{4}$	44⁴⁄₉%
5/9	$0.55\overline{5}$	55⁵⁄₉%
7/9	$0.77\overline{7}$	77⁷⁄₉%
8/9	$0.88\overline{8}$	88⁸⁄₉%
1/10	0.1	10%
1/12	$0.083\overline{3}$	8⅓%
1/16	0.0625	6¼%

7. Percentages Properly speaking, a percent is a ratio expressed as a number followed by a percent sign. 50% is a percent, and not a percentage. A *percentage* is what results when you take a percent *of* something. When bank robbers gather to divide up the loot, each gets a percentage of the take; if Lefty took half the risk, his percentage should be 50% of the money. Although the words are often used interchangeably, the distinction is a useful one, as the next section will show.

WORKING WITH PERCENTS

1. Finding a Percent To find what percent one number is of another, divide the percentage by the base.

◆ After completing a year of employment, a clerk's salary of $32,000 is increased by $1600. The base salary is $32,000 and the raise (which is a percentage of the original salary) is $1600. Thus the percent increase is $1600 \div 32,000 = \frac{1}{20} = \frac{5}{100} = 5\%$.

◆ If the cost of an airline ticket rises from $200 to $250, the increase is $50. The *percent increase* is $50 \div 200 = \frac{25}{100} = 25\%$. Note that the choice for the base is the original value of $200—that is, the cost prior to any increase or decrease.

Correctly identifying the base is the key to avoiding the most common misuse of percents. Note, for example, that 100% of anything is the entire amount. 100% of $200 is $200; thus if $200 is reduced by 100%, it becomes 0. (If an advertisement announced that prices had been slashed 100%, it would mean that they were giving away the store, which would be highly unlikely.)

2. Solving Percent Problems Almost all problems involving percents take one of the three forms given below:

   i. What is 40% of 500?
   ii. 12 is what percent of 480?
   iii. 60% of what number is 900?

In each instance, three quantities—the *rate*, the *base*, and the *percentage*—are related by a simple formula.

$$\text{rate} \times \text{base} = \text{percentage}$$

The *rate* is the stated or implied percent, which is always a percent *of* something; that "something" is called the *base*. The result of taking a percent of something is the *percentage*. Finally, the word *of* implies multiplication, and the word *is* translates into an equal sign.

   i. "What is 40% of 500?" asks for the percentage: _____ = 40% × 500.
   ii. "12 is what percent of 48?" asks for the rate: 12 = _____ × 48.
   iii. "60% of what number is 900?" asks for the base: 60% × _____ = 900.

These three questions correspond to the three possible ways of writing the percent formula given above:

   i. rate × base = percentage

   ii. rate = $\dfrac{\text{percentage}}{\text{base}}$

   iii. base = $\dfrac{\text{percentage}}{\text{rate}}$

The solutions to the above examples are:

   i. Find the percentage: 40% × 500 = 0.40 × 500 = 200.

   ii. Find the rate: $\dfrac{12}{48} = \dfrac{1}{4} = 25\%$.

   iii. Find the base: $\dfrac{900}{60\%} = \dfrac{900}{0.60} = \dfrac{9000}{6} = 1500$.

◆ In qualifying for a mortgage, a borrower must meet income requirements that are expressed in percent form. Specifically, the monthly cost of the new home must not exceed 28% of the prospective buyer's gross monthly income. If the house would cost $900 each month (including property tax and insurance), then 28% of the gross monthly income should be at least $900. What is the lowest qualifying gross monthly income? This is equivalent to asking: 28% of what number is 900, or 28% × _____ = 900? Using equation (c) above, this can be rewritten as: $^{900}\!/_{.28}$ = 3214.29. Multiplying the monthly income by 12 gives the qualifying annual income: $38,571.48.

◆ In pari-mutuel betting, a racetrack takes 15% of all the money wagered on a particular race, and distributes the rest as winnings. If $78,000 is bet on a particular race, find the track's percentage.

    This translates into an equation of type (a), in which the rate and base are known. The answer is 0.15 × 78,000, or $11,700.

PERCENTILES

Percentiles come up most often in situations in which people, places, or things are ranked in order of size, ability, speed, income level, or any other numerically measurable attribute. The *percentile rank* is a way of referring to those in the group who come out on top. For example, the 95th percentile identifies those that rank in the top 5%.

◆ A group of 100 people is arranged in order of age from youngest to oldest. The oldest 25% (the last 25 people in the line) are referred to as the *75th percentile* because 75% of the group is younger.

The *50th percentile* divides an ordered group in half. The term used to describe the cutoff point of the 50th percentile is the *median* (see also Chapter 5). In a group of scores ordered from lowest to highest, half fall above the median and half below. The 50th percentile of the group refers to the upper half.

The 75th percentile may also be referred to as the top *quartile*. In any sample of test scores—the SAT, GRE, MCAT, or LSAT exams, for example—there are four quartiles, which divide all scores into a top 25%, a bottom 25%, and two middle 25% groups (which are called the *interquartile range*). Often a distribution of test scores is divided into *stanines*. Although the word may sound exotic, a stanine is simply one of nine equal divisions of 11.1%, and can be thought of as a rating from 1 to 9. The median would fall in the center of the fifth stanine.

If an ordered group is divided into fifths, each fifth is referred to as a *quintile*. A group broken into quintiles consists of five equal divisions of 20% each.

♦ A commentator reports that between 1980 and 1989 federal tax rates paid by the bottom three quintiles of the income distribution rose, while tax rates declined for those in the top two quintiles. This statement divides the taxpaying population into the bottom ⅗ and the top ⅖, which together constitute the entire population. The statement says nothing about actual amounts; instead it refers to *relative* amounts: ⅗ (or 60%) of the whole group versus ⅖ (or 40%) of the group.

Beyond quartiles, quintiles, and stanines, the most commonly used division of ordered scores or numerical data is a *decile*, which refers to 10 equal divisions of 10%.

APPLICATIONS OF DECIMALS AND PERCENTS

1. Points and Basis Points In financial markets, percentages are referred to as *points*. For home mortgages, a point is prepaid interest in the amount of 1% of the amount borrowed. In the stock and bond markets, a *basis point* refers to ¹⁄₁₀₀ of 1% of the yield (or interest rate of earnings) of an investment. 100 basis points equals 1%; thus one basis point equals ¹⁄₁₀₀ of 1%.

♦ If the yield of a 30-year Treasury bond climbed 25 basis points in one day's trading, then the interest rate rose ²⁵⁄₁₀₀%, or ¼ of a percent.

2. Majority and Plurality The terms *majority* and *plurality* arise most often in connection with political elections and polls and with demographic studies. The majority of a group is any subgroup that constitutes more than 50% (or half) of the group. In a vote or study in which people divide into two groups, the larger group is the *majority*. However, if a population is divided into more than two groups, none of which accounts for more than half of the total (meaning that none constitutes a majority), the largest group is called a *plurality*. A plurality can also refer to the amount by which the largest group exceeds the next largest group in size.

• In the 1992 presidential election Bill Clinton received 42.95% of the popular vote, followed by 37.4% for George Bush, 18.86% for Ross Perot, and less than 1% for other candidates. Presidents are not elected by popular vote (in fact two presidents—Hayes and Harrison—were elected despite losing the popular vote), but in this case the plurality voted for Clinton. Only 55.9% of eligible voters actually cast votes in the election. Thus President Clinton received the votes of about 24% of the voting-age population (because 42.95% of 55.9% equals 0.4259×0.559, which is 0.24, or 24%—a plurality, but not a majority).

3. Tipping Tips for service are calculated either as flat fees (for valet parking or coat checking) or as a percentage of a bill. A standard tip is 15% of the pretax total for good service, and 20% of the bill for exceptional service. In most situations, tips must be computed mentally and quickly, and therefore certain rules of thumb can be useful.

In states where the meal tax is 5%, triple the amount of the tax listed on the check in order to compute a 15% tip. Unfortunately, this method will result in a disappointingly low tip in states where the alcohol tax is already included in the menu price—that is, where the meal tax is based on the price of the food but not on the drinks. An alternative is to compute 10% of the pretax total (a simple matter of moving the decimal point one place to the left) and add half of that amount again.

• You want to add a 15% tip to the following restaurant check:

Food and drink	$31.55
Tax (on food only)	1.53
Total	$33.08

First round the pretax total upward to $32. 10% of $32 is $3.20, to which half is added: $3.20 + 1.60 = 4.80. The tip should be about $5. Since the check totals just over $33, it would be reasonable to leave $38. In each of these calculations, the totals are rounded *up*, which errs on the side of generosity.

4. Sales Tax/Meal Tax The taxability of various types of purchases differs from state to state. While food purchased in grocery stores is usually not taxed, food served in restaurants generally is (except where listed prices already include the tax). Non-food items and prepared food sold in supermarkets are usually subject to state sales tax. Consider the following grocery-store receipt:

Cookies	$ 2.49
Milk	1.12
Eggs	0.99
Apples	2.53
Wax paper	1.98
Liquid detergent	3.49
Subtotal	12.60
Tax	0.27
Total	12.87

The tax amounts to 5% of the cost of the non-food items: $0.05 \times (1.98 + 3.49) = 0.2735$, which rounds to 27 cents.

A 5% sales tax or meal tax is typical, but the percentage varies from state to state and from country to country. As a benchmark, 5% can easily be calculated as half of 10%. Thus 5% of $25.89, for example, is half of 2.589, or about $1.30.

5. Discounts A sale sign will often make a claim such as "UP TO 60% OFF." In theory, this means that as much as 60% (or ⅗) of a tagged price will be deducted from that price at the register (before sales tax is computed) for selected items. This deduction is called a *discount*.

    ♦ A pair of shoes that lists for $145 is advertised at 60% off. The discount is 60% of $145, or $0.60 \times 145 = \$87.00$. Thus the cost of the shoes will be $145 - 87 = \$58$, plus tax if applicable.

A more direct way to compute the sale price (as opposed to the amount of the discount) is to subtract the percent discount from 100 and multiply the result by the list price. Thus a 30% discount (30% off) is equivalent to paying 70% of the list price.

    ♦ A $65 shirt is advertised at 30% off. Thus it will cost 70% of $65, which is $45.50.

6. Markups A *markup* is a fraction (or percentage) of a base price that is added onto the base price. Markups are generally not advertised as enthusiastically as discounts. They arise in the case of difficult-to-obtain items (such as popular car models) or products that require a considerable amount of extra service to deliver to the market. A markup usually occurs in the sale by wholesaler to retailer, and again in the sale by retailer to consumer. But some retailers will mark up a suggested retail price yet again if the demand is sufficient.

♦ A popular sports car is priced for the national market at $15,500 with no options. Because of demand, a local dealer marks this price up 6%. The amount of the markup is 6% of 15,500, or 0.06 × 15,500 = $930. Thus the local market price is $15,500 + 930 = $16,430. This represents (100 + 6)%, or 106%, of the original base price.

Φ ——————————————————————————— Φ

SHIFTING BASES AND DRIFTING SNOWS

A 25% markup followed by a 25% markdown does not restore the original price. This is because the base amount changes after the markup.

Suppose a $1000 snowblower is marked up 25% so that its price increases by $250, to become $1250. When winter ends the price is marked down 25%. But 25% of $1250 is 0.25 × 1250 = $312.50. The adjusted price is less than the original price of $1000. It works out to $1250 − 312.50 = $937.50. In other words, because of the shifting value of the base, a 25% increase is not canceled out by a 25% decrease. Similarly, a 25% decrease in price followed by a 25% increase does not restore the original value, again because of the shifting base. In fact the same result is obtained: $937.50.

Φ ——————————————————————————— Φ

7. Inflation The rate of inflation reflects the buying power of one dollar, which tends to diminish with the passage of time. The process is not irreversible, but the prevailing trend is for prices to steadily rise over time. (The word *inflation* is meant to describe a long-term trend. A seasonal trend such as the rising price of gasoline in the summer months is not considered inflationary because the price declines to normal levels in the fall.)

The amount of the average increase in prices is marked each year by a number referred to as an *index*. In its mathematical sense, an index is a percent, ratio, or reference number that is used to measure the growth or decline of a quantity over a given time period. In the case of inflation, the U.S. Department of Labor regularly calculates a *cost-of-living index* known as the *Consumer Price Index* (or CPI). This index reflects changes in the market prices of a wide range of commodities (food, clothing, and household supplies), and thus measures the cost of living for an average consumer. See Table 1.3.

The CPI, like all indexes, consists of numbers that have no intrinsic meaning, but are used to indicate *percent change over time*. This type of change is called *relative change*.

• Between 1950 and 1990 the Consumer Price Index jumped from 24.1 to 130.7—an increase of 106.6. To find the percent increase, divide the size of the increase by the CPI in 1950. That is,

percent increase = 106.6 ÷ 24.1 = 4.42, or 442%

• The price of a half-gallon of orange juice increases 10 cents in one year, from $1.79 to $1.89. The relative change can be calculated as the change in price divided by the original price. Here the original price is 179 cents, so the relative increase is $^{10}/_{179}$, which is about 0.056, or 5.6%.

Table 1.3. Consumer Price Indexes

Year	CPI	Year	CPI	Year	CPI
1950	24.1	1966	32.4	1981	90.9
1951	26.0	1967	33.4	1982	96.5
1952	26.5	1968	34.8	1983	99.6
1953	26.7	1969	36.7	1984	103.9
1954	26.9	1970	38.9	1985	107.6
1955	26.8	1971	40.5	1986	109.6
1956	27.2	1972	41.8	1987	113.6
1957	28.1	1073	44.4	1988	118.3
1958	28.9	1974	49.3	1989	124.0
1959	29.1	1975	53.8	1990	130.7
1960	29.6	1976	56.9	1991	136.2
1961	29.9	1977	60.6	1992	140.3
1962	30.2	1978	65.3	1993	144.5
1963	30.6	1979	72.5	1994	148.2
1964	31.0	1980	82.4	1995	152.4
1965	31.5				

An inflation rate of 4% per year means that an item priced at $100 at the beginning of the year would cost $104 at year's end. This would correspond to an index value of 100 at the beginning of the year, and 104 at year's end.

Φ ——————————————————————————— Φ

A CASE OF RAMPANT INFLATION?

In the fall of 1759 a Philadelphia merchant named Thomas Willig ordered a custom-built card table from the finest cabinet shop in the city. The finished product exceeded all expectations. Marking a high point of the Chippendale style, it featured ball-and-claw feet with exquisitely carved vines and leaves twisting up the legs. And yet when Willig's executors itemized his possessions after his death in 1821, they merely noted "one mahogany card table," to which they assigned the value 50 cents.

In 1964 Willig's great-great-great-grandson discovered his ancestor's furniture in a bank vault, and he took the lot to be appraised. The table, perhaps the most precious item, made its way to this descendant's home, where it remained until 1991. It was then consigned to Sotheby's auction house, where it sold for the princely sum of $950,000. A question that immediately comes to mind is, What rate of inflation does this represent?

The surprising answer is less than 9% per year—about what the Dow Jones average has sustained during this century. Instead of being an example of astronomical inflation, the Chippendale table's increase in value is actually quite modest. As the future-value formula on page 000 shows, a sum of 50 cents invested for 170 years (1821 to 1991) at 9% would be worth well over a million dollars, even if compounded just once a year. Such is the nature of exponential growth.

Φ ——————————————————————————— Φ

8. Constant Dollars Increments to salaries and hourly wages are usually measured against the rate of inflation. If a salary does not increase by at least the same percentage as the inflation rate, then the employee's buying power goes down—he cannot buy as much as he could the previous year. The use of the terms *real dollars, real wages,* and *constant dol-*

lars reflects the fact that the buying power of a dollar changes over time, and therefore comparing today's dollars to last year's dollars requires an unchanging unit of measure. Thus "dollars" are fixed to a particular time, and are identified as "today's dollars" or "1920 dollars." The term *current dollars* refers to the face value of the amount that changes hands in a transaction.

Suppose the newspaper reports that "Since 1989, the real wages of high school dropouts have declined by 20%." Such a statement is about buying power, and not about actual numbers. It could be that the average wage of high school dropouts remained constant in those years, while inflation rose 20% since 1989.

Table 1.4 shows the buying power of a dollar from 1950 to 1995. The 1950 column shows that the buying power of a 1950 dollar had declined to 16 cents in 1995. That is, after 45 years under a mattress, a 1950 dollar would purchase only 16% of what it could have purchased in 1950. The numbers in parentheses are multipliers. The bottom row, for example, shows that a 1950 dollar had 6.38 times the buying power of a 1995 dollar, a 1960 dollar had 5.18 times the buying power, and so on.

Table 1.4. The Value of a Current Dollar

Year	1950 $	1960 $	1970 $	1980 $	1990 $
1950	1	1.23 (.81)	1.61 (.62)	3.42 (.29)	5.42 (.18)
1955	.90 (1.11)	1.11 (.90)	1.45 (.69)	3.07 (.33)	4.87 (.21)
1960	.81 (1.23)	1	1.31 (.76)	2.78 (.36)	4.40 (.23)
1965	.76 (1.31)	.94 (1.06)	1.23 (.81)	2.61 (.38)	4.13 (.24)
1970	.62 (1.61)	.76 (1.31)	1	2.12 (.47)	3.36 (.30)
1975	.45 (2.23)	.55 (1.81)	.72 (1.38)	1.53 (.65)	2.43 (.41)
1980	.29 (3.42)	.36 (2.78)	.47 (2.12)	1	1.59 (.63)
1985	.22 (4.47)	.28 (3.63)	.36 (2.77)	.76 (1.31)	1.21 (.83)
1990	.18 (5.42)	.23 (2.54)	.30 (3.36)	.63 (1.59)	1
1995	.16 (6.38)	.19 (5.18)	.25 (3.95)	.54 (1.87)	.85 (1.17)

Another way to use the table is to compare salaries. To find the 1995 equivalent of a $20,000 annual salary in 1970, read down the *1970 $* column to the year 1995. The multiplier shown is 3.95. Multiplying $20,000 by 3.95 gives the equivalent 1995 annual salary of $79,000.

The values in the table are based on the Consumer Price Index.

9. After-Tax Earnings Earnings on stock-market investments, as well as interest earned on bank accounts, constitute taxable income (unless the money is tax-sheltered, as explained in Chapter 6), so taxes will reduce the earnings. To calculate after-tax earnings as a percent of the amount invested, use this formula:

$$\text{rate of return (after taxes)} = i \times (1 - t)$$

where i is the annual percent interest earned (or annual yield), and t is the tax rate.

+ An investor in a 28% tax bracket earns 15% in one year in a mutual fund. The after-tax rate of return is

$$15\% \times (1 - 28\%) = 0.15 \times (1 - 0.28) = 0.108 = 10.8\%$$

Rounding Off

A number may be rounded off whenever it is more precise than is necessary for its intended purpose, or when its accuracy is in doubt, as in a measurement. To understand how numbers are rounded, it is useful to define the idea of a *significant digit*.

1. The "significance" of a digit has to do with its own reliability and the reliability of the digits that precede it. In a rough measurement of 240 meters, for example, the 2 is very reliable and the 4 is reliable to the extent that we know we are not dealing with 230 or 250 meters. But we could be dealing with something as big as 245 or as small as 235. The zero, then, is not reliable at all. Instead of the word "reliable," scientists use "significant."

In general, any nonzero digits in a number are significant. Zeros that fall between nonzero digits are also significant, and zeros that fall at the end of a decimal can be significant if they establish the accuracy of the number. The examples below show how this works:

+ 0.5 is numerically equivalent to 0.50 or 0.500. But as a measurement, 0.5 meters is not interchangeable with 0.50 meters or 0.500 meters. The first is accurate only to the tenths place, the second is accurate to the hundredths place, the last to the thousandths place. So 0.5 has one significant digit, 0.50 has two significant digits, and 0.500 has three significant digits. Notice that the first zero in each case is not a significant digit.

+ 17.4 has three significant digits. Therefore, as a measurement it is closer to 17.4 than to either 17.3 or 17.5.

+ 17.40 has four significant digits. Thus it is closer to 17.40 than to either 17.39 or 17.41.

◆ 0.0174 has three significant digits.

◆ 0.01074 has four significant digits.

In a whole number, zeros at the end may or may not be significant, depending on the context.

> ◆ A crowd at a demonstration was estimated to be 15,000 people. This figure has only two significant digits.

> ◆ The paid attendance at a baseball game was *exactly* 23,700. Because this figure is understood to be exact, the zeros are significant. The figure thus has five significant digits.

Each significant digit establishes the degree of accuracy of the preceding decimal place. Thus the first 0 in 1,503,000 is significant because its own degree of accuracy is established by the 3 that follows it. However, the first 0 in 1,500,000 may or may not be a significant digit.

2. Truncating and Rounding Down The simplest method of rounding off is by *truncation*—that is, shortening. This is usually done to the part of a decimal number that lies to the right of the decimal point. Truncating a number means eliminating all of the significant digits beyond a certain place value. In the case of a mixed number, truncation involves eliminating the fractional part of the number.

> ◆ If a scientific calculator truncates numbers that have more than 10 significant digits, a repeating decimal such as 0.12 would appear as 0.1212121212.

> ◆ The number $5\frac{2}{3}$ can be truncated to 5 by eliminating the fraction.

The process of truncating at or to the left of the decimal point is referred to as *rounding down*. In rounding down, the last significant digits are replaced by zeros.

> ◆ 75,648 can be rounded down to 75,000.

> ◆ Payoffs on racetrack bets are rounded down to the nearest multiple of 10 cents. Thus a winning $2 ticket at 5-to-3 odds should pay $3.33, but the track rounds this down to $3.30. The 3-cent difference is referred to as *breakage*. (For more on horse racing, see Chapter 5.)

3. Rounding to the Nearest In any measurement, the last significant digit is the least certain digit because the digit that would follow is unknown. In general, a significant digit is used to round to the decimal place that precedes it. There are several ways this can be done. The most common method is called *rounding to the nearest*—that is, rounding to the nearest designated place value. If a digit is followed by 5, 6, 7, 8, or 9, it is rounded up; if it is followed by 0, 1, 2, 3, or 4, it remains the same.

- An interest payment calculated by a bank as 15.23758 dollars must be rounded to the nearest cent. In this instance, the 3 in the cents position is followed by a 7, which indicates that the 3 should be rounded up to a 4, making the interest payment $15.24.

- If rounded to the nearest dollar, 15.23758 would become 15, because the 5 in the ones place is followed by a 2. If rounded to the nearest dime it becomes $15.20; to the nearest cent it becomes $15.24.

4. Rounding to the Nearest Even Digit When the last significant digit is 5, it is arbitrary whether to round the previous digit up or down. The standard practice is to round up all 5's. But the practice favored by scientists is to *round to the nearest even digit*. This assures that in a list of numbers, half of such decisions result in rounding up and the other half in rounding down.

- 645 rounded to the nearest even digit would be 640, whereas 375 rounded to the nearest even digit would be 380.

Number Sequences

Certain sequences of numbers, because of their usefulness or familiarity, have acquired names and reputations. The odd numbers (1, 3, 5, 7, 9, . . .) and the even numbers (0, 2, 4, 6, 8, 10, . . .) are the most commonly encountered sequences. Both of these are examples of *arithmetic sequences* (also known as *arithmetic progressions*).

1. An *arithmetic sequence* is a list of numbers in which each number (or *term*) is derived from the preceding number by adding a constant value to it. That is, there is a common difference between any two successive terms of an arithmetic sequence.

- The common difference in the sequence of even numbers is 2.

Φ ———————————————————————— Φ

WHEN TO ROUND OFF MEASUREMENTS

All measurements have round-off error because no measuring device is exact. A meter stick, for example, has markings for millimeters, and these allow accuracy to within 0.5 mm. A yardstick or tape measure is accurate to $\frac{1}{16}''$. Consequently, when measurements are combined, the least accurate measurement should dictate the accuracy of the result.

When adding measurements: All measurements should be rounded to the accuracy of the least precise measurement. (For example, the sum 8.6 cm + 0.14 cm + 2.75 cm should be reformulated as 8.6 cm + 0.1 cm + 2.8 cm. The result is 11.5 cm.)

When multiplying or dividing measurements: A product or quotient is rounded to the number of significant digits of the measurement with the fewest significant digits. (For example, 6.2 m × 8.75 m = 54.25 m^2. Because 6.2 has two significant digits, this result should also be rounded to two significant digits, leaving 54 m^2.)

Φ ———————————————————————— Φ

♦ The sequence 5, 10, 15, 20, 25, . . . is an arithmetic sequence with a common difference of 5.

♦ The sequence 3.5, 7, 10.5, 14, 17.5, . . . is an arithmetic sequence with a common difference of 3.5.

The sum of an arithmetic sequence is given by the formula $S = n \times (a + b) \div 2$, where S is the sum, n is the number of terms, and a and b are the first and last terms. If the number of terms is an odd number, the sum is the middle term multiplied by the number of terms.

♦ $1 + 3 + 5 + 7 + 9 + 11 = 6 \times (1 + 11) \div 2 = 36$

2. A *geometric sequence* is a list of numbers in which each term is derived from the previous term by multiplying by a fixed constant value (called a *ratio*). A common geometric sequence is: 1, 2, 4, 8, 16, 32, 64, . . . , in which the ratio is 2.

♦ The sequence 1, $\frac{2}{3}$, $\frac{4}{9}$, $\frac{8}{27}$, $\frac{16}{81}$, . . . is a geometric sequence. Its ratio is $\frac{2}{3}$.

Φ ——————————————————————— Φ

THERE IT LIES!

At the age of 10, Carl Gauss, who would go on to become one of the most prolific mathematical geniuses ever, was simply one of many boys taking a class in arithmetic. His teacher, intending to put everyone to work for a while, told the boys to sum all of the counting numbers from 1 to 100, and they set to work. Within a few seconds, Gauss stood up with his slate, walked to the teacher's desk, laid it down (as was the custom for students who had finished a long problem), and announced, "There it lies!" The teacher was incredulous. Gauss had written the number 5050, and nothing else.

Asked to explain himself, Gauss said he noticed that the sum of the first and last numbers is 101 (1 + 100), and that each pair working in from the outside also summed to 101—that is, 2 + 99, 3 + 98, 4 + 97, all the way to 50 + 51. Seeing that there were 50 such pairs, he multiplied 50 by 101 to get 5050. This reasoning, which can be applied to any arithmetic sequence, leads to a general formula for such a sum: $S = n \times (a + b) \div 2$, where a is the first term, b the last, and n the number of terms.

Φ ——————————————————————— Φ

3. The sum of the first n terms of a geometric sequence with first term a and common ratio r is given by

$$\text{sum} = \frac{a \times (r^n - 1)}{r - 1}$$

* The sum of 3, 6, 12, 24, 48, 96, 192, 384—a geometric sequence of 8 terms, in which the first term is 3 and the common ratio is 2—is given by

$$\text{sum} = \frac{3 \times (2^8 - 1)}{2 - 1} = 765$$

This formula is the basis for the annuity formulas given in Chapter 5.

4. The Population Principle Arithmetic and geometric sequences figure in the celebrated *population principle* outlined by Thomas Malthus (1766–1834) in the first two editions of his *Essay on the Principle of Population* (1798 and 1803). This essay established the starting point of all later discussions of population growth. Its main argument rests on a mathematical point: Malthus's famous conclusion that "population, when unchecked, grows in a geometrical ratio" while "subsistence [i.e., food] only increases in an arithmetical ratio." As Malthus confidently concluded, "a slight acquaintance with numbers will show the immensity of the first power in comparison of the second."

What Malthus means by an "arithmetical ratio" is an arithmetic sequence such as 1, 2, 3, 4, 5, ... By a "geometrical ratio" he means a geometric sequence such as 1, 2, 4, 8, 16, . . . , in which the next number is derived by multiplying the previous number by the same value. Here the multiplier is 2. With population growing so rapidly compared to the rate of food production, Malthus could only envision an outcome of "misery and vice." Fortunately, although Malthus's assessment of population growth was essentially accurate, his predictions about the potential of agricultural production were not. Unfortunately, the immediacy of the problem Malthus perceived in 1800 has only been delayed. His statistics may have been off, but his essential mathematical argument, if somewhat oversimplified, is still valid.

Table 1.5 shows Malthus's own population growth table, in which the percent increase each year is not always the same (which would be the case in a true geometric progression) but close enough to 5% per year to qualify as an example of a "geometrical ratio."

Table 1.5. The Population of England, from Malthus's *Essay* (7th ed.)

Year	Population	Increase	Percent Increase (per 5 years)
1780	7,721,000		
1785	7,998,000	277,000	3.59%
1790	8,415,000	417,000	5.21%
1795	8,831,000	416,000	4.94%
1800	9,287,000	456,000	5.16%
1805	9,837,000	550,000	5.92%
1810	10,488,000	651,000	6.62%

Malthus estimated the doubling time of the population to be 83⅓ years.

5. The Fibonacci Sequence A famous sequence of numbers begins with two 1's and generates each new term as the sum of the previous two. It is called the *Fibonacci sequence,* and it looks like this:

1, 1, 2, 3, 5, 8, 13, 21, 34, 55, 89, . . .

Notice that every number on the list (after the first two) is the sum of the two numbers that precede it. The sequence is named for the thirteenth-century mathematician Leonardo Fibonacci of Pisa, and it is more than a mathematical curiosity. It is frequently discussed in studies of population growth patterns. Fibonacci himself came upon it as the solution to the following problem: How many pairs of rabbits can be produced from a single pair of rabbits in one year if each pair produces a new pair each month, but only after having matured to reproductive age over an initial two-month period? The growth pattern is shown below. Note how the first pair does not produce a new pair until the third month, but then produces a new pair each month thereafter. The second pair does not produce a new pair until the fifth month, and so on.

Month:	1	2	3	4	5	6	7	8	9	10	11	12
No. of pairs:	1	1	2	3	5	8	13	21	34	55	89	144

This sequence also describes such biological patterns as the distribution of leaves spiraling around a lengthening stem, the distribution of scales on a pinecone, and the distribution of seeds in a sunflower.

Infinity and Infinitesimal

One of the most important and subtle concepts in all of mathematics is the idea of *infinity*. The symbol for infinity (∞), while often thought of as the largest number, is not a number at all. It is best thought of as the numerical process of increasing perpetually. Thus any infinitely large quantity is one that cannot be contained within any boundary and is not exceeded by any given number. A quantity that is "infinitely small," or *infinitesimal*, is smaller than any number or dimension that can be named.

These concepts are necessary in order to describe the real number line and the real number system. Because the number line has no end in either direction, it is infinite. Thus there are an infinite number of real numbers, an infinite number of counting numbers, an infinite number of odd numbers, and so on.

The idea of infinity, in the sense of "infinitely large," leads naturally to the idea of the infinitesimal, or "infinitely small." This is accomplished with *reciprocals*. The reciprocal of 10 is $\frac{1}{10}$, the reciprocal of 100 is $\frac{1}{100}$, the reciprocal of 1000 is $\frac{1}{1000}$. The greater the number, the smaller its reciprocal. (Small, in this case, means closer to zero.) As the counting numbers count off to infinity, their reciprocals move closer and closer to zero on the real number line. Infinitely close, but not equal.

This notion of the infinitesimal is what allows us to talk about approaching a fixed value and getting "infinitely close" to it. The sum

0.3 + 0.03 + 0.003 + 0.0003 + 0.00003 + . . . is a good example; adding these figures gives the following sequence of numbers:

    0.3
    0.33
    0.333
    0.3333
    0.33333
    and so on.

The "end" result, if we could keep on adding indefinitely, would be the repeating decimal 0.3333 . . . , which is equal to ⅓. Not only are there an infinite number of counting numbers, but there are an infinite number of real numbers between any two counting numbers. In that sense the interval between 0 and 1 is infinite. To convince yourself of this, simply divide the interval in half (at ½), then divide the halves in half (at ¼ and ¾), then the fourths in half, and so on. The very phrase "and so on" is perhaps the best evidence of the presence of the infinite in everyday life.

The Vocabulary of Arithmetic

Scientists love to name things, so it should come as no surprise that there are technical terms that describe the parts of an addition, subtraction, multiplication, or division of two numbers. Many of these terms are considered the outworn vocabulary of old arithmetic primers. Yet they have not disappeared from the language, because they describe things that have no other proper names. Here is a rundown of their meanings:

1. When two numbers or quantities are added together, the first number is referred to as the *augend* and the second as the *addend*. The result of the addition is called a *sum*.

$$15 + 7 = 22$$

augend   addend   sum

2. In a subtraction, the number being subtracted from is called the *minuend*, and the number being subtracted is called the *subtrahend*. The result of any subtraction is called a *difference*.

$$8 - 5 = 3$$

minuend   subtrahend   difference

Φ ——————————————————————————— Φ

DOES 0.$\overline{9}$ EQUAL 1?

Perhaps the most reassuring of all numbers is 1. It divides evenly into any number. It is easy to multiply by. So it is particularly unsettling when someone tries to pass off 0.$\overline{9}$ as 1. Sports fans could not be expected to say, "We're number point nine repeating!" with much conviction. Nor can we commend a lover who says "You're my point nine repeating and only." Although the implications of the number 1 are many and profound, when regarded merely as a signpost on the road from 0 to 2 it is no more special than one-third, which we all agree is equal to "point three repeating," or to two-thirds, which is surely "point six, dot dot dot" (see page 7 for a demonstration of this). In this progression, 1 can be thought of as three-thirds, which is clearly "point nine repeating."

$$\frac{1}{3} = 0.3333 \ldots$$

$$\frac{2}{3} = 0.6666 \ldots$$

$$\frac{3}{3} = 0.9999 \ldots = 1$$

Φ ——————————————————————————— Φ

3. The result of a multiplication is called a *product*. The numbers that are multiplied to get the product are referred to as the *multiplicand* and the *multiplier.*

$$9 \times 15 = 135$$

multiplicand multiplier   product

When two positive integers are multiplied together, each is called a *factor* of the product. Because $3 \times 7 = 21$, we say that 3 and 7 are *factors* of 21. The only other factors of 21 are the numbers 1 and 21 itself.

4. The words that describe the parts of a division are still current. The *dividend* is the number being divided, the *divisor* is the number doing the dividing, and the *quotient* is the result of dividend divided by divisor. In long or short division, the quotient might include a *remainder.*

$$\overset{\frown}{}32 \div 4 = 8\overset{\frown}{}$$
$$\underset{\text{dividend}}{} \quad \underset{\text{divisor}}{} \quad \underset{\text{quotient}}{}$$

$$45 \div 6 = \tfrac{45}{6} = 7\tfrac{3}{6} = 7\tfrac{1}{2}$$
$$\underset{\text{dividend}}{} \quad \underset{\text{divisor}}{} \quad \underset{\text{quotient}}{} \quad \underset{\text{remainder}}{}$$

Arithmetic Operations

The four basic arithmetic operations—addition, subtraction, multiplication, and division—are associated with the four symbols $+$, $-$, $\times$, $\div$. Although these are the standard symbols seen on simple four-function calculators and in elementary textbooks, they are not the only symbols used to represent these operations. Moreover, the notation can be ambiguous—many symbols can be used to indicate multiplication and division. And the plus and minus signs have alternative meanings as well—a fact which can and does cause confusion.

1. The Addition Sign: $+$ This symbol is used to indicate the summation of two or more quantities. Order is unimportant. Thus $a + b$ is the same as $b + a$. Even if three or more numbers are to be added together, the way in which they are combined is arbitrary. Thus $a + b + c$ can be calculated as the sum $(a + b)$ added to c, or as the number a added to the sum $(b + c)$.

The word used to describe the interchangeability in the order of an operation is *commutativity*. Addition is said to be commutative because $a + b = b + a$.

The property describing the arbitrariness of grouping is called *associativity*. Addition is said to be associative because $(a + b) + c = a + (b + c)$, where the operations within parentheses are carried out first.

- $4 + 7 + 3$ can be written as $(4 + 7) + 3$ or as $4 + (7 + 3)$. The first sum is $11 + 3$, whereas the second is $4 + 10$. In both cases the final sum is the same.

The plus sign ($+$) can also indicate direction or position. For example, $+\infty$ ("plus infinity") represents an infinitely large positive number. In the context of stock price quotations, a plus sign means an increase; the figure $+2\frac{1}{2}$ indicates a net gain of $2.50 in a stock's value during the day's trading. In general, the plus sign, when not used to indicate the operation of addition, implies a positive or increasing value.

2. The Subtraction Sign: − This symbol indicates the difference between two quantities. In using this symbol, order is important. In other words, the operation of subtraction is *not* commutative: $a − b$ does not equal $b − a$. Subtraction is also *not* associative, which means that the grouping of subtracted quantities is not arbitrary.

    ♦ The compound difference $35 − 14 − 6$ is meant to be read from left to right. First subtract the 14 from 35 to get 21, then subtract 6. Thus $35 − 14 − 6 = (35 − 14) − 6 = 21 − 6 = 15$. Notice that if 6 is subtracted from 14 first, the result would be $35 − (14 − 6)$, or $35 − 8$, or 27, which is incorrect.

Subtraction can be thought of as a shift along the number line. The subtraction sign, like the plus sign, can indicate direction.

    ♦ $22 − 11$ indicates a position that is 11 units to the *left* of 22 on a number line. The result is 11.

```
                    11 units
    ┌──────────────────────────────────────┐
    ↤
    10 11  12  13  14  15  16  17  18  19  20  21  22  23
```

    ♦ $5 − 12$ indicates a position that is 12 units to the left of 5 on a number line. The result is $−7$.

```
                    12 units
      ┌────────────────────────────────────┐
      ↤
    −8 −7 −6 −5 −4 −3 −2 −1  0  1  2  3  4  5  6
```

3. The last example indicates another use of this symbol. When it appears before an isolated number or expression, it indicates an *additive inverse*. The additive inverse of a given number is another number which, when added to the first number, gives the value 0. Thus the additive inverse of a positive number is a negative number, and vice versa. The additive inverse of 9 is $−9$, and the additive inverse of $−14$ is 14.

4. Like the plus symbol, the minus symbol is used to indicate direction. In engineering, a $−5\%$ grade indicates a downward inclination of a road or a pipe; the drop is 5 meters in every 100 meters of horizontal travel.

Φ ———————————————————————————— Φ

MINUS VS. NEGATIVE

Among the more commonly confused mathematical terms, *minus* and *negative* have the dubious honor of confounding teachers as well as students. The word *minus* refers to the operation of subtraction, not to negative numbers. As a part of speech, it is a preposition meaning "diminished by." It is not a verb, and should never be used as one. 7 minus 4 equals 3; but you cannot "minus" 4 from 7. More importantly, "minus 4" is not interchangeable with "negative 4." The word *negative* is an adjective used to describe additive inverses of positive numbers. In other words, the "negative" numbers are those that lie to the left of 0 on the number line. They are not called "minus" numbers. Thus it is technically incorrect to describe the symbol −3 as "minus 3"—"negative 3" is the proper form.

Other overused mathematical prepositions include *plus*, which means "increased by," and *times*, meaning "multiplied by." Like *minus*, *plus* is often improperly used to identify position on the number line. "Plus 5" is a poor substitute for "positive 5." As for *times*, its use as a verb has become the error of choice among many high school and college students. "You times it by 7" will never become standard usage if it is constantly corrected—which it should be, even if (and especially if) you find your teacher saying it.

Φ ———————————————————————————— Φ

5. The *additive inverse* of any number is its mirror image on the number line with respect to 0. Thus the additive inverse of 5 is −5, and the additive inverse of −³⁄₂ is ³⁄₂. The sum of any number and its additive inverse is 0.

ADDITION AND SUBTRACTION

1. **Adding and Subtracting Negative Numbers** Adding a negative number is equivalent to subtracting its additive inverse.

- 5 + (−9) is equivalent to 5 − 9, which equals −4.
- −13 + −7 is equivalent to −13 − 7, which equals −20.

Subtracting a negative number is equivalent to adding its additive inverse.

♦ 24 − (−14) is equivalent to 24 + 14, which equals 38.

2. Adding Fractions with Common Denominators Two or more fractions that are to be added or subtracted are said to have *common* denominators when their denominators are the same. To add two or more fractions with common denominators, add the numerators and place this sum over the common denominator.

♦ $\dfrac{2}{9} + \dfrac{5}{9} = \dfrac{2+5}{9} = \dfrac{7}{9}$

♦ $\dfrac{2}{15} + \dfrac{7}{15} + \dfrac{4}{15} = \dfrac{2+7+4}{15} = \dfrac{13}{15}$

3. Subtracting Fractions with Common Denominators To subtract fractions with common denominators, subtract one numerator from the other, and retain the common denominator.

♦ $\dfrac{11}{12} - \dfrac{7}{12} = \dfrac{11-7}{12} = \dfrac{4}{12}, \text{ or } \dfrac{1}{3}$

4. Lowest Common Denominators The *lowest common denominator* (LCD) is the easiest choice of common denominator to work with. It is the smallest number that is divisible by both denominators. If the denominators have no common factor (other than 1), the LCD is their product.

♦ The LCD of ⅕ and ⅓ is 5 × 3 because 5 and 3 have no common factors.

If the denominators do share some common factors, follow this procedure to find the LCD:

  i. Write the denominators in a row, leaving spaces in between them.
  ii. Find any prime number (other than 1) that divides evenly into one or more of the denominators.
  iii. Divide each denominator by the prime number, and write the result below each number that divides evenly. Carry down those denominators that are not divisible by the prime number.

iv. Repeat the process until the bottom row contains all 1's.

v. Multiply all of the prime divisors. Their product is the LCD.

♦ Find the LCD of $\frac{1}{12}$, $\frac{2}{9}$, and $\frac{1}{18}$.

List the denominators:	12	9	18
Divide each by 2, carrying down the numbers not divisible by 2:	6	9	9
Now repeat the process, dividing by 3:	2	3	3
Repeat, dividing by 3 again:	2	1	1
Finally, divide by 2:	1	1	1

The LCD is the product of the divisors: $2 \times 3 \times 3 \times 2 = 36$.

5. Adding and Subtracting Fractions with Different Denominators To add or subtract two fractions with different denominators, first find the LCD. Then convert both fractions to equivalent fractions having the LCD as their denominator. Converting a fraction into an equivalent fraction with a particular denominator requires multiplying the numerator and denominator by the same number.

♦ $\frac{2}{3} + \frac{4}{5}$ converts to $\frac{2}{3} \times \frac{5}{5} + \frac{4}{5} \times \frac{3}{3}$, or $\frac{10}{15} + \frac{12}{15}$, which equals $\frac{22}{15}$.

♦ $\frac{3}{4} - \frac{11}{7}$ converts to $\frac{3}{4} \times \frac{7}{7} - \frac{11}{7} \times \frac{4}{4}$, or $\frac{21}{28} - \frac{44}{28}$, which equals $\frac{-23}{28}$.

MULTIPLICATION AND DIVISION

1. The Multiplication Sign ×, *, · There are many ways to indicate the multiplication of two numbers or quantities. The best known of these, the cross, is perhaps the least used in practice. Far more common is the dot (·), or even the asterisk (*) used in computer programming. The most common indicator of multiplication is no symbol at all. Multiplication will be symbolized in the following text by the cross, the dot, and the absence of any symbol.

♦ The circumference of a circle whose radius is r units is given by the expression $2\pi r$, which means $2 \times \pi \times r$.

2. Multiplication is nothing more than repeated addition. The sum $a + a + a + a$ can be expressed as $4 \times a$, $4 * a$, $4 \cdot a$, or simply as $4a$, and a sum such as $4 + 4 + 4 + 4 + 4$ is more briefly expressed as 5×4, $5 * 4$, or $5 \cdot 4$.

3. Like addition, multiplication is both *commutative* and *associative*. This means that $a \times b$ is equal to $b \times a$. It also means that when multiplying three or more numbers, the way in which the numbers are grouped does not affect the final outcome. $5 \times 4 \times 3$ can be thought of as $(5 \times 4) \times 3$ or as $5 \times (4 \times 3)$—the result is the same.

Although $5 \times a$ does equal $a \times 5$, the product is written as $5a$ rather than $a5$. In general: *Numbers that multiply letters are written first.*

4. The Division Symbol: ÷ Like the multiplication symbol $\times$, the symbol ÷ for division is rarely encountered in everyday settings. A fraction is generally preferred as a means of indicating a quotient of two values. Thus $12 \div 4$ usually appears as $12/4$, $12/4$, or $\dfrac{12}{4}$.

5. Division is the opposite (or inverse) of multiplication. Consequently, every multiplication can be written in the form of a division, and vice versa.

- 25×3 is equivalent to $25 \div \dfrac{1}{3}$.

- $12 \div 4$ is equivalent to $12 \times \dfrac{1}{4}$.

6. The *multiplicative inverse,* or *reciprocal,* of any number other than 0 is found by dividing that number into 1. Consequently, the product of any number and its multiplicative inverse is 1.

- The multiplicative inverse of 5 is $\dfrac{1}{5}$. Thus $5 \times \left(\dfrac{1}{5} \right) = 1$.

The reciprocal of any counting number is 1 over that number.

- The reciprocal of 12 is $\dfrac{1}{12}$.

- The reciprocal of -10 is $\dfrac{1}{-10}$, which is the same as $\dfrac{-1}{10}$.

The reciprocal of a simple fraction can be found by flipping it.

- The reciprocal of $\frac{2}{3}$ is $\frac{3}{2}$.

- The reciprocal of $\frac{-4}{5}$ is $\frac{5}{-4}$, which is the same as $\frac{-5}{4}$.

The reciprocal of a mixed number can be found by first converting it to an improper fraction and then flipping it.

- The mixed number $5\frac{2}{3}$ is equivalent to $\frac{17}{3}$. Its reciprocal is $\frac{3}{17}$.

Φ ————————————————————————————————— Φ

DIVISION BY ZERO

While zero divided by any nonzero number equals zero ($\frac{0}{5} = 0$, for example, because there are 0 5's in 0), division *by* zero is not defined. Thus $\frac{5}{0}$ is not defined and is not considered a number. This follows from the rules of algebra, and also explains why the expression $\frac{0}{0}$ does not equal zero. It is instead referred to as an *indeterminate form*.

Φ ————————————————————————————————— Φ

7. Multiplying Fractions To multiply two simple fractions, multiply their numerators and denominators separately. To multiply two mixed numbers, first convert to improper fractions, then multiply numerators and denominators. To multiply a whole number and a fraction, multiply the numerator by the whole number.

- $\frac{2}{5} \times \frac{7}{6} = \frac{2 \times 7}{5 \times 6} = \frac{14}{30}$, which reduces to $\frac{7}{15}$.

- The product of $3\frac{1}{4}$ and $5\frac{2}{3}$ is the same as the product of $\frac{13}{4}$ and $\frac{17}{3}$. The result is $\frac{13 \times 17}{4 \times 3}$ or $\frac{221}{12}$, which can be expressed as $18\frac{5}{12}$.

- $5 \times \frac{3}{7} = \frac{5 \times 3}{7} = \frac{15}{7}$.

8. Dividing by a Fraction Dividing by a fraction is the same as multiplying by its reciprocal.

♦ $2 \div \dfrac{1}{4} = 2 \times 4 = 8$

9. To facilitate multiplying or dividing an integer by a fraction or a fraction by an integer, the integer can be expressed in fractional form by using 1 as its denominator. Thus 5 can be written as ⁵⁄₁.

♦ $\dfrac{7}{8} \div 5$ is equivalent to $\dfrac{7}{8} \div \dfrac{5}{1}$, or $\dfrac{7}{8} \times \dfrac{1}{5}$, which equals $\dfrac{7 \times 1}{8 \times 5}$, or $\dfrac{7}{40}$.

Φ ———————————————————————— Φ

HOW MANY PINTS IN A GALLON?

The idea of dividing by a fraction is more challenging than the idea of dividing by a whole number. Dividing by 2, for example, means dividing into two equal parts. Dividing by 5 means dividing into five equal parts. So how are we supposed to interpret dividing by ¼?

It helps to think of division as an operation that determines how many times one number *is contained in* a second number. For instance, $2 \div \frac{1}{4}$ asks how many fourths are contained in two units. One way to visualize this is by picturing how many quarts there are in two gallons. That is, dividing two gallons into quarter-gallons (or quarts) is the same as dividing 2 by ¼. There are 4 quarts in a gallon, and thus 2×4 or 8 quarts in 2 gallons. The same idea can be conveyed by asking how many quarters there are in two dollars.

Φ ———————————————————————— Φ

10. Multiplication Shortcuts: Cancellation One way to multiply fractions faster while reducing the chance of errors is to "cancel" common factors in numerators and denominators. If any numerator shares a factor with any denominator in a multiplication, each should be divided by that factor, with the result written in their places. If one of the numerators is the same as one of the denominators, they may be crossed out (or "canceled"), leaving a 1 in each of their places.

$$\bullet \quad \frac{7}{8} \times \frac{4}{9} \times \frac{1}{7} = \frac{1}{2} \times \frac{1}{9} \times \frac{1}{1} = \frac{1}{18}$$

11. Multiplication Shortcuts: The Distributive Law With the advent of affordable pocket-sized calculators, the need to mentally compute or estimate sums, differences, products, and quotients has diminished, although it has not disappeared. The ability to calculate mentally is largely a matter of practice, but being acquainted with certain basic rules, one of which is the *distributive law,* can help greatly.

In symbols, the *distributive law of multiplication over addition* says the following:

$$a \cdot (b + c) = a \cdot b + a \cdot c$$

and

$$(a + b) \cdot (c + d) = a \cdot (c + d) + b \cdot (c + d) = a \cdot c + a \cdot d + b \cdot c + b \cdot d$$

The utility of this rule relies on the ease of multiplying by powers of 10, by multiples of 10, or by multiples of 5.

- The product 7×15 can be thought of as $7 \times (10 + 5)$, or 7 *tens* plus 7 *fives.* In symbols, this can be written as

$$7 \times 15 = (7 \times 10) + (7 \times 5) = 70 + 35 = 105$$

DIVISIBILITY

Searching for factors of a given whole number is a matter of investigating its *divisibility.* One number is divisible by another when dividing the first number by the second leaves no remainder. Any even number, for example, is divisible by 2; thus if a whole number ends in 0, 2, 4, 6, or 8, it is divisible by 2. This is the simplest divisibility test. Here's another: a number is divisible by 3 if the sum of its digits is divisible by 3.

These and other divisibility tests are summarized below in Table 1.6.

Table 1.6. Divisibility Tests

A whole no. is divisible by	if
2	its last digit is 0, 2, 4, 6, 8.
3	its digits sum to a number divisible by 3. (If the sum of the digits is too large to determine divisibility by 3, sum the digits of the sum of the digits and check again. Repeat the process until it is clear whether or not the resulting number is divisible by 3.)
4	its last two digits, taken together as a single number, are divisible by 4.
5	its last digit is 0 or 5.
6	it passes the tests for divisibility by 2 and 3.
7	when the ones digit is split off, doubled, and subtracted from the number that remains, the remainder is divisible by 7. (This process may be repeated until divisibility by 7 is established or ruled out.)
8	its last three digits, taken as a single number, are divisible by 8.
9	the sum of the digits is divisible by 9. (As with 3, the sum of the digits of the sum of the digits may have to be found. This process can be continued until a single digit remains; if it is 9, the original number is divisible by 9.)
10	the last digit is 0.
11	if the difference between the sum of the 1st, 3rd, 5th, etc., digits and the sum of the 2nd, 4th, 6th, etc., digits is 0 or 11.
12	it passes the tests for divisibility by 3 and 4.
7, 11, 13, 17, 19	(See the appendix for an elaborate test for these divisors.)

The following examples illustrate several of the divisibility tests.

Divisibility by 4: 743,328 is divisible by 4 because 28 is divisible by 4.

Divisibility by 7: 546 is divisible by 7 because it splits into 54 and 6. Double the 6 and subtract from 54: $54 - (2 \times 6) = 42$, which is divisible by 7. 1617 splits into 161 and 7. Double the 7 and subtract: $161 - 14 = 147$. If you don't recognize this as a multiple of 7, split the 14 and 7, double the 7, and subtract to get 0, which is divisible by 7.

Divisibility by 8: 743,328 is divisible by 8 because 328 is divisible by 8.

Divisibility by 9: 743,328 is divisible by 9 because $(7 + 4 + 3 + 3 + 2 + 8)$ is divisible by 9.

Divisibility by 11: 8415 is divisible by 11 because $(8 + 1) - (4 + 5) = 0$. 41,943 is divisible by 11 because $(4 + 9 + 3) - (1 + 4) = 11$.

Φ ——————————————————————————— Φ

FATE HAS NOT SMILED UPON YOU

Three tax collectors, Messrs. A, B, and C, came to the door of a lowly old scrivener to ask for a tithe. "I'd rather not," said the scrivener. But they refused to take no for an answer, so the scrivener changed his tune.

"Well, then," he said. "Would each of you please write down a three-digit number."

A wrote 247, B wrote 638, and C wrote 914.

"I'll tell you what," said the old man. "Let's make it a six-digit number by writing your number twice."

So A wrote 247247, B wrote 638638, and C wrote 914914.

"In lieu of my debt," said the scrivener, "I will pay a number of gold pieces to Mr. A equal to the remainder when he divides his number by 7, to Mr. B equal to the remainder when he divides his number by 11, and to Mr. C equal to the remainder when he divides his number by 13. Will you accept this offer?"

The three conferred and quickly calculated that with divisors like 7, 11, and 13, they could get as much as 6 + 10 + 12, or 28 gold pieces, and they were sure to get at least half of that, which was still greater than the debt. They eagerly accepted. But after a flurry of calculating, they found that their gain amounted to nothing.

"Take heart, gentlemen," said the scrivener, "it is merely bad luck." Why don't we change divisors. Let Mr. A divide by 13, Mr. B by 7, and Mr. C by 11." And the collectors went right to work only to discover that, alas, they came up with nothing again. The scrivener sympathized. "Why don't you change divisors once more," he said, "and surely something will come to you." But again— nothing, nothing, nothing.

"Gentlemen," said the scrivener, "my sympathies to you, but you chose the numbers yourselves, and fate has not smiled upon you. Good day."

And with that he closed his door behind him, leaving three sad men to ponder how Lady Luck could have been so unkind to them three times.

The story is adapted from Karl Menninger's *Calculator's Cunning*, an interesting book about mental arithmetic. It is a well-known example of a losing proposition.

As you might have guessed, fate had nothing to do with the outcome. Any six-digit number of the form described in the story—three digits written twice—is divisible by 1001 (try to figure out why). And 1001, it turns out, is the product of 7, 11, and 13.

Φ ———————————————————————————— Φ

FACTORING

1. When the only factors of a number are 1 and the number itself, the number is called a *prime number*. Examples of prime numbers include 2, 3, 5, 7, 11, 13, 17, 19, and 23.

2. Every counting number can be expressed as a product of prime numbers in exactly one way. This is called a *prime factorization*. The prime factorization of any number is found by splitting it into factors, then splitting its factors into factors until all of the factors are primes. Below are a few examples of this process.

- $42 = 6 \cdot 7 = 2 \cdot 3 \cdot 7$
- $315 = 3 \cdot 105 = 3 \cdot 5 \cdot 21 = 3 \cdot 3 \cdot 5 \cdot 7$
- $780 = 2 \cdot 390 = 2 \cdot 2 \cdot 195 = 2 \cdot 2 \cdot 5 \cdot 39 = 2 \cdot 2 \cdot 3 \cdot 5 \cdot 13$

Factor trees are the most efficient way to factor large numbers. Each branching splits a factor into its own factors, and branches end with primes. The answer is found at the branch ends, when all the primes are gathered together. Here is an example, illustrating how 5460 can be factored into $2 \cdot 2 \cdot 3 \cdot 5 \cdot 7 \cdot 13$.

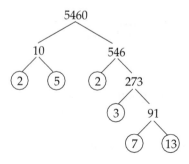

Factor tree

Table 1.7. Prime Numbers up to 150

2	41	97
3	43	101
5	47	103
7	53	107
11	59	109
13	61	113
17	67	127
19	71	131
23	73	137
29	79	139
31	83	149
37	89	

Exponents and Roots

1. Just as repeated addition of the same number can be replaced by multiplication, repeated multiplication of the same number can be facilitated by *exponentiation.*

- The product $3 \times 3 \times 3 \times 3 \times 3$ can be written 3^5.

- $7 \cdot 7 \cdot 7 \cdot 7 \cdot 7 \cdot 7 \cdot 7 \cdot 7 \cdot 7 \cdot 7 = 7^{10}$

In these examples, the superscript is called an *exponent* or a *power.* (In some books it is referred to as the *index.*) The multiplied number is referred to as the *base.* The expression 3^5 can be read as "three to the fifth," which is short for "three raised to the fifth power."

2. Exponents are the basis of *scientific notation,* in which a given number is expressed as a number between 1 and 10 multiplied by a power of 10. (For more on scientific notation, see page 54.) A positive power of 10 in decimal form is a 1 followed by the number of zeros indicated by the exponent. 10^5 is written in decimal form as a 1 followed by five zeros, or 100,000. A billion is a 1 followed by nine zeros; it can be written instead as 10^9.

3. Any nonzero number raised to the zero power equals 1. Also, the first power of any number is the number itself.

- $10^0 = 1$
- $e^0 = 1$
- $5^1 = 5$
- $10^1 = 10$

Φ ——————————————————————————— Φ

CASTING OUT THE 9'S

The test for divisibility by 9 involves summing the digits and checking to see if the result is divisible by 9. If the number so derived is too big, its own digits may be summed, and the process may be carried to its extreme, in which case a single number will remain. This number is called the *9-remainder.* If the 9-remainder is 9, then the original number, as well as all of the sums found along the way, are divisible by 9. If the number is not 9, it will represent the remainder when the original number is divided by 9 (thus the term "9-remainder"). For example, start with the number 843,797,183,256. The sum of the digits is 63, and the sum of these digits is 9. Thus the number is divisible by 9.

This may seem like a lot of work. So you might want to try a shortcut called *casting out the 9's.* Here's how it works: Before summing the digits, cross out any 9's, as well as any pairs or triplets of digits that sum to 9. Try this with: 843,797,183,256. Working from left to right, cross off 8 and 1, 4 and 5, 3 and 6, 7 and 2, and 9. All that remains is 7, 8, and 3, which sum to 18, which sum to 9. By the same method, you should be able to see at a glance that if 36,547 is divided by 9, the remainder is 7.

This leads directly to a useful rule of thumb for balancing the books. In any sum of numbers, a transposition error—one that involves reversing the order of digits—creates a discrepancy that is divisible by 9. This is because a transposition does not change the sum of the digits, or the 9-remainder, and the difference between any two numbers with the same 9-remainder must be divisible by 9. For example, the numbers 37 and 73 differ by 36, which is divisible by 9. The numbers 327 and 732 (whose 9-remainders are both 3) differ by 405, which is divisible by 9. (Recall that a number is divisible by 9 if its digits sum to a number that is divisible by 9.)

Φ ——————————————————————————— Φ

---

4. The second power is often referred to as the *square* of a number. The third power may be referred to as the *cube.* Thus 10^2 can be read as "ten squared" or "the square of ten," and 10^3 as "ten cubed" or "the cube of ten."

5. A *negative exponent* expresses a *reciprocal* of a power of the base. That is, a base raised to a negative power can be rewritten as 1 over the same base raised to the positive power.

- 10^{-3} means $\dfrac{1}{10^3}$ or $\dfrac{1}{10 \cdot 10 \cdot 10}$ or $\dfrac{1}{1000}$.

- 2^{-4} means $\dfrac{1}{2^4}$ or $\dfrac{1}{2 \cdot 2 \cdot 2 \cdot 2}$ or $\dfrac{1}{16}$.

6. Negative powers of 10 are converted to decimal form by starting with 1 and moving the decimal point to the left the number of places given by the negative exponent.

- 10^{-1} equals 0.1. (That is, start with 1.0 and move the decimal point one place to the left. Note that the result is the equivalent of $\frac{1}{10}$, which is the meaning of 10^{-1}.)

- 10^{-2} equals $\dfrac{1}{10^2}$, or 0.01.

- 10^{-5} is the same as 0.00001.

7. To multiply a whole number by a positive power of 10, write zeros after the number, the number of zeros being the same as the exponent.

- 35×10^2 is the same as 3,500.
- 4×10^6 is the same as 4,000,000.

To multiply any decimal number by a power of 10, move the decimal point to the right or left the number of places indicated by the exponent. If the exponent is positive, move the decimal point to the right; if negative, to the left.

- 7.5×10^3 is the same as 7,500.
- 4.75×10^{-3} is equivalent to 0.00475.
- 7×10^{-4} is equivalent to 0.0007.

8. Another form of exponentiation is the process of taking roots. The *square root* of a number is a number which, when multiplied by itself (or squared), equals the original number. A *cube root* of a number, when cubed, equals the number.

- The square roots of 16 are 4 and −4, because $4^2 = 16$ and $(-4)^2 = 16$.
- The cube root of 125 is 5 because $5^3 = 125$.

9. The notation used to indicate a root is the *radical sign* $\sqrt{}$. An expression containing a radical sign is called a *radical*. Whatever appears underneath a radical sign is referred to as a *radicand*. In the expression $\sqrt{25x^2}$ the radicand is $25x^2$.

By itself, the radical sign indicates a positive square root, which is also known as the *principal square root*. Thus $\sqrt{16}$ equals +4. It is also true that −4 is a square root of 16, because the product of two negative numbers is positive; thus $(-4) \times (-4) = 16$. This is conveyed in radical form using a minus sign, as in $-\sqrt{16} = -4$. To indicate both the positive and negative square roots of 16, the symbol $\pm\sqrt{16}$ is used.

10. The symbol for a cube root is $\sqrt[3]{}$. Thus $\sqrt[3]{125} = 5$. The symbol for a fourth root is $\sqrt[4]{}$, for a fifth root $\sqrt[5]{}$, and so on. $\sqrt[4]{81} = 3$ because $3^4 = 81$.

11. Roots can also be represented by *fractional exponents*. This avoids the use of radical signs altogether. The root is indicated by the denominator of the fractional exponent.

- The expression $4^{1/2}$ means $\sqrt[2]{4}$ or $\sqrt{4}$.
- The expression $125^{1/3}$ means $\sqrt[3]{125}$.

12. When a fractional exponent has a numerator other than 1, it connotes both a power and a root. In the expression $8^{2/3}$, the 2 indicates a second power (or square) and the 3 indicates a third root (or cube root). These can be taken in either order—root first and then power, or power first and then root. Thus the expression $8^{2/3}$ can be written and calculated in two ways:

$$8^{2/3} = \sqrt[3]{8^2} = \sqrt[3]{64} = 4$$

$$\text{or}$$

$$8^{2/3} = \left(\sqrt[3]{8}\right)^2 = (2)^2 = 4$$

13. Beware of negative radicands where the root is even. Square roots of negative numbers, for example, do not exist among the real numbers. Neither do fourth or sixth roots of negative numbers. Such roots lead to the realm of *imaginary numbers* and *complex numbers*. Both are built upon the concept of the imaginary number *i*, which is defined as the square root of -1. Thus

- $\sqrt{-1} = i$
- $\sqrt{-4} = \sqrt{4} \cdot \sqrt{-1} = 2i$
- $\sqrt{-25} = 5i$

Logarithms

1. An *exponential* is an expression that consists of a base and an exponent (or power). For example, the exponential 5^3 has a base of 5, an exponent of 3, and a value of 125. In this scheme, the word *logarithm* refers to the exponent. The logarithm of 125 is 3 because 3 is the power of 5 that gives the value 125. Thus a logarithm *is* an exponent. But to make sense a logarithm must refer to a base. In the above example, 3 would be the "base-5 logarithm" of 125. This is designated $\log_5 125$. A logarithm can be defined using any positive number except 1 for the base. Thus there are base-5 logarithms, base-2 logarithms, base-7 logarithms, and so on, but in practice only two bases are used.

2. Base-10 logarithms, called *common logarithms,* are implied by the simple abbreviation *log*. Any scientific calculator will compute base-10 logarithms by means of a key marked *log* or *logx*. *The common logarithm of a given number is the power to which 10 must be raised in order to equal that number.*

- Because $10^2 = 100$, the common logarithm of 100 is 2.
- The common logarithm of 100,000 is 5 because $10^5 = 100,000$.
- The common logarithm of 0.01 is -2 because $10^{-2} = 0.01$.

Numbers that are not integer powers of 10 also have common logarithms, although they are irrational numbers and in practice are always rounded off. The value of log2, for example, is given by a calculator as 0.30103, which means that $10^{0.30103} = 2$ (approximately, but close enough).

3. A *natural logarithm* uses the number *e* as its base. Like π, *e* is an irrational number (a nonterminating, norepeating decimal) that occurs in many physical applications. Its value is approximately 2.718. The natural logarithm is designated *ln* in textbooks and on calculators. *The natu-*

ral logarithm of a number is the power to which e *must be raised in order to equal the number.*

- Using a calculator, the natural logarithm of 5 comes up as 1.609437912. This means that *e* raised to the power of 1.609437912 equals 5. (Again, the value of the logarithm is not exact due to rounding.)

4. The term *antilog* is still used occasionally to refer to the process of raising 10 or *e* to a given power. Thus an antilog is an exponential expression, usually of base 10, but sometimes of base *e*. Taking 2 as a common logarithm, the antilog would be 10^2, or 100. Antilogs were more common in the days of slide rules, when logarithms greatly facilitated computations that are now routinely done with handheld calculators. You entered the realm of logarithms by finding the log on a slide rule, and you left it by finding the antilog.

LOGARITHMIC SCALES

Logarithms are a convenient way of dealing with numbers that cover a very wide range. This is because logarithms are exponents. Therefore, a scale of numbers that runs from a value as small as 0.01 to one as large as 1,000,000 would run only from −2 to 6 if it were expressed using common logarithms (since 0.01 is 10^{-2} and 1,000,000 is 10^6).

The examples below all use such logarithmic scales. The first illustrates the use of a logarithmic number line in place of a standard number line as the basis of a graph. The other examples, such as the Richter scale, the pH scale, and the scale for star magnitudes, show how something assigned a value of 4 is not twice the size of something with a value of 2. This is because units on any logarithmic scale represent *ratios* of relative size. The 4 might mean 10^4, and the 2 might mean 10^2, in which case the first is 100 times the size of the second. (In addition to the examples below, another logarithmic scale—the decibel scale—is explained in Chapter 4.)

1. Kepler's Third Law The figure below illustrates Johannes Kepler's famous Third Law of Planetary Motion, which relates each planet's distance from the sun to the time it takes to complete an orbit around the sun. The data is contained in Table 1.8, although Kepler's Law is hidden in the numbers. The relationship emerges when the orbital times are squared and the distances from the sun are cubed. Another way to see the relationship is to compare the logarithms of the two numbers for each planet. This is what the graph does.

The figure is an example of a logarithmic graph. Notice that the labels on the horizontal and vertical axes are not the expected values 1, 2, 3, etc., but instead are powers of 10. Thus the intervals represent a compression of time and distance, and the result is a straight-line graph. The law as Kepler stated it is best understood in the context of ratios and proportions (which are explained in detail in Chapter 2). The law says: The square of a planet's solar orbit time is proportional to the cube of its mean distance from the sun. Table 1.8 gives orbital times in terms of earth years, and distances in terms of the earth's mean distance from the sun, which is defined as 1 *astronomical unit* (AU). (To see how astronomical units came about, see Chapter 4.)

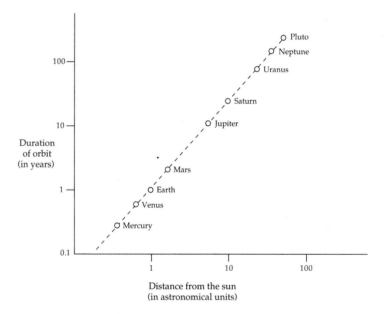

In the following table, all measures are based on the earth's time of orbit and distance from the sun. To verify Kepler's law, square any one of the periods and cube the corresponding distance.

Table 1.8. Planetary Orbital Periods and Distance from the Sun

	Mercury	Venus	Earth	Mars	Jupiter	Saturn	Uranus	Neptune	Pluto
orbital period (in years)	0.24	0.62	1	1.88	11.86	29.46	84	165	248
distance (in AU's)	0.39	0.72	1	1.52	5.20	9.54	19.18	30.06	39.44

2. The Richter Scale The Richter scale was developed by Charles F. Richter and Beno Gutenberg of the California Institute of Technology in 1935 to measure the magnitude of earthquakes. A logarithmic scale, it begins at a magnitude of 0, which in Richter's time was undetectable. With each unit increase, the Richter scale represents a *tenfold increase* in the size of the earthquake. Thus a quake measuring 2 on the scale is ten times greater than one measuring 1. A quake of 8.0 on the scale is not twice as powerful as one measuring 4.0, but rather 10^4 or 10,000 times as powerful. The scale, then, is logarithmic because it shows the relative sizes of quakes in terms of powers of 10.

The Richter scale is based on measurements collected by a device called a *seismograph,* which is located (ideally) 100 kilometers (or 62 miles) from the epicenter of the quake. Although the scale has no upper limit, the largest earthquakes measured have not exceeded 9.0 on the scale. A quake measuring 2.0 or below is barely perceptible to an observer at the epicenter. Any quake over 7.0 is considered major, although the amount of physical damage and loss of life depend largely upon the stability of buildings and the geological composition of the region. Table 1.9 shows the magnitudes of some of the most damaging earthquakes. Those that predate the Richter scale are estimates.

Table 1.9. Earthquake Magnitudes (Richter Scale)

Place	Year	Magnitude
Los Angeles	1994	6.6
Japan	1993	7.8
Iran	1990	7.7
San Francisco	1989	7.1
Armenia	1988	6.9
Mexico City	1985	8.1
Great Britain	1984	5.5
Alaska	1964	8.4
Chile	1960	8.3
Soviet Union	1952	8.5
San Francisco	1906	8.3
Colombia	1906	8.6
Krakatau	1883	9.9

To calculate the relative magnitude of two earthquakes, divide the larger Richter-scale magnitude by the smaller and use the resulting value as a power of 10.

• Comparing San Francisco's 1906 and 1989 quakes, find the difference in magnitudes: $8.3 - 7.1 = 1.2$. Then find the antilog of this number: $10^{1.2} = 15.85$. Thus the 1906 quake was almost 16 times as powerful as the 1989 quake.

3. The pH Scale The pH scale was devised by the Danish chemist S. P. L. Sørensen in 1909 as a measure of the acidity or alkalinity of any solution. It is a logarithmic scale that runs from 0 to 14. Numbers less than 7 are in the *acidic* range; numbers greater than 7 are in the *alkaline* range. "pH" means the *p*ower of *H*ydrogen ion concentration. The scale is logarithmic in that it gives the concentration of ions in the form of reciprocals of powers of 10. A high pH value corresponds to a large negative power of 10, meaning a weaker concentration of hydrogen ions. A low pH value indicates a small negative power of 10, which represents a higher concentration.

• Pure water has a concentration of 10^{-7} hydrogen ions per liter. This translates into a pH factor of 7, which is neither acidic nor alkaline.

• Vinegar has a concentration of 10^{-4} hydrogen ions per liter. Therefore its pH value is 4, which is acidic.

• Hydrochloric acid has a pH value of 0.1; sodium hydroxide has a pH value of 14. These represent the outermost extremes of the scale.

Because it is logarithmic, the pH scale indicates a tenfold change in concentration for each unit increment or decrement.

• Lemon juice has a pH of about 2.1, whereas a fruit jelly might have a pH of 3.1. Because the pH difference is 1, and this represents a power of 10, the juice is ten times more acidic than the jelly, which means that it has ten times the concentration of hydrogen ions.

• Rainwater with a pH of 4 is ten times as acidic as rainwater with a pH of 5. The pH level of rain is of crucial importance to farmers because soil fertility is directly related to its acidity. Acidic soils are less hospitable to crops than to conifers, and are usually treated with lime to "sweeten" the soil, or reduce its acidity.

Table 1.10 shows pH values for a variety of common substances.

Table 1.10. pH Values

Acidic	pH
Sulfuric acid	0.1
Orange juice	3.0
Wine	3.4
Tomatoes	4.2
Coffee	5.0
Tap water	5.8
Milk	6.9
Basic	
Blood	7.4
Seawater	8.2
Baking soda	8.5
Milk of magnesia	10.0
Household ammonia	11.9
Lye	13.0
Potassium hydroxide	14.0

4. Magnitude of Stars (Celestial Magnitude) There are two scales that measure the brightness or visibility of stars. Each is logarithmic, since orders of magnitude correspond to ratios of 2.5. The brighter the star, the lower the number assigned to it. Thus a star of magnitude 1 is 2.5 times brighter than a star of magnitude 2, which is 2.5 times brighter than a star of magnitude 3, and so on.

The number 2.5 is not exact. It is an approximation of $\sqrt[5]{100}$, which is close to 2.51188643. Thus a star of magnitude 0 is 100 times brighter than a star of magnitude 5.

The scale is arbitrary, but it has a historical basis—the star catalog of the Greek astronomer Hipparchus (c. 150 B.C.), which classifies about 850 visible stars according to their brightness. Hipparchus assigned magnitudes from 1 to 6, grouping the brightest stars as magnitude 1 and the faintest as magnitude 6.

In 1850 the English astronomer N. R. Pogson refined the system using the logarithmic scale outlined above. Because of improvements in astronomical observations, the scale expanded to include decimal values, as well as negative values to accommodate the brightest stars. In Pogson's scale the sun has a celestial magnitude of -26.8, which means that it is about 2.5^{27} (or 55 billion) times brighter than the brightest star. But this is only from the perspective of the earth. The magnitude of stars as viewed from the earth is called *apparent magnitude*. On this scale a 6 represents the magnitude of the faintest stars that can be seen with the unaided eye, and between 0 and 1 represents the magnitude of the brightest visible stars, Sirius and Arcturus.

For scientific purposes, it is necessary to have a relative scale of magnitude that can be used to compare the brightness of stars in an absolute sense. This is called *absolute magnitude*. In theory it is based on the brightness of stars when viewed from a distance of 10 parsecs. (A parsec is equivalent to 3.259 light-years; it is defined more precisely in Chapter 4.) The absolute magnitude of a star can be calculated from its apparent magnitude using a formula that involves its distance (*d*) from earth. The formula adjusts the apparent magnitude by a factor related to the square of its distance from the viewer. If *M* is the absolute magnitude and *m* is the apparent magnitude, then

$$M = m - 5\log(d) + 5$$

Table 1.11 shows the relation between these variables for several celestial objects.

Table 1.11. Celestial Magnitudes

Name	m	M	Distance from earth (in parsecs)
Moon	−11	32	1.25×10^{-8}
Sun	−26.7	4.7	4.85×10^{-6}
Sirius	−1.43	1.42	2.65
Arcturus	−0.06	−0.2	11
Antares	1.00	−5.1	162
Betelgeuse	0.5	−6.0	180
Rigel	0.08	−6.3	185
Canopus	−0.73	−4.5	200
Deneb	1.26	−7.5	429

Although it appears from earth as merely one of many bright stars, Deneb, in the Cygnus constellation, has an absolute magnitude of −7.5, which is 12.2 orders of magnitude greater than the sun. Because orders of magnitude are logarithmic and represent powers of 2.5, Deneb gives off more than $2.5^{12.2}$—or 70,000—times as much light as the sun. Estimates of celestial distances are imprecise (especially those of Canopus and Rigel), so the absolute magnitude should be thought of as a rough measure.

Factorials

1. A *factorial* is an operation that applies to the counting numbers. It is represented by an exclamation point, which indicates the product of any counting number with all of the counting numbers below it.

• "Five factorial," written 5!, means $5 \times 4 \times 3 \times 2 \times 1$, which equals 120.

• "Seven factorial" is written 7! and equals $7 \times 6 \times 5 \times 4 \times 3 \times 2 \times 1$, or 5,040.

• By definition, "zero factorial," or 0!, is equal to 1.

2. Factorials are used to count the number of ways that a set of distinct objects can be rearranged. Such rearrangements are called *permutations*. (For more information on permutations and games of chance, see Chapter 5.)

• The number of permutations of the three letters A, B, and C is 3!, which is equal to $3 \times 2 \times 1$, or 6. (They are ABC, ACB, BAC, BCA, CAB, and CBA.)

• The number of ways of arranging 10 people in a line is 10 factorial. There are 10 possible candidates for the first position, 9 for the second, 8 for the third, and so on. The total number of possible permutations is the product of the number of possible candidates for each position: $10 \times 9 \times 8 \times 7 \times 6 \times 5 \times 4 \times 3 \times 2 \times 1$. Thus there are 3,628,800 possible permutations of 10 people.

3. Most scientific calculators have a special key for finding factorials. It is denoted by an exclamation point, or by $x!$.

Order of Operations

1. Unlike reading, arithmetic is not always carried out from left to right (or even from right to left). In the most basic ordering of operations, multiplication and division come before addition and subtraction. In the expression $8 + 7 \times 5$, therefore, the multiplication must be performed first. It sometimes helps to translate such expressions into words: $8 + 7 \times 5$ means "8 plus *the product of* 7 and 5," which simplifies to 8 plus 35, or 43.

2. In the absence of any *grouping symbols* (parentheses, brackets, radical signs), an expression involving only addition and subtraction is calculated from left to right. Thus $10 - 7 + 14$ is calculated as the difference $10 - 7$ added to 14. The result is 17.

3. When multiplication and division appear in the same expression with no parentheses or other grouping symbols, the operations are carried out from left to right.

• In the expression $20 \div 5 \times 3$, the division is carried out first: $20 \div 5$ is 4, which is then multiplied by 3 to arrive at 12.

4. Any operations contained within parentheses must be carried out before operations that are outside of the parentheses.

◆ $(7 - 10) \times (6 + 5) = (-3) \times (11) = -33$

5. Expressions such as powers, roots, factorials, or logarithms should be calculated before any other algebraic operations.

◆ 2×10^2 equals 2×100 (and not 20^2).

◆ $5 + 3!$ equals $5 + (3 \times 2 \times 1) = 5 + 6 = 11$ (not 8!).

◆ $4 \cdot \log 100$ equals $4 \cdot 2$, or 8 (since $\log 100 = 2$).

6. The initials PPMDAS, which stand for "Powers and Parentheses, Multiplication and Division, Addition and Subtraction," can help you remember the order of operations. With this in mind, the best way to proceed with arithmetic computations is to calculate first the values of any expressions that are contained within grouping symbols (parentheses, brackets, or radical signs). If an expression contains nested parentheses (or other grouping symbols), calculation should begin with the innermost set of parentheses. Within parentheses or among grouped expressions, the order PPMDAS should be followed.

◆ $4 \times [100(2 \times 3)^2] = 4 \times [100(6)^2] = 4 \times [100 \times 36] =$ $4 \times [3600] = 14,400$

7. The fraction bar is also a grouping symbol. Thus, in the expression $\dfrac{5}{25 - 10}$ the subtraction in the denominator is carried out first, followed by the division.

◆ $\dfrac{5}{25 - 10} = \dfrac{5}{15} = \dfrac{1}{3}$

A sequence of operations written without grouping symbols can be confusing. This is particularly true when division is indicated using the slash in place of a fraction bar. Expressions such as $1/4 - 3$ or $7 + 3/20$ would be much clearer if written with parentheses as $(1/4) - 3$ and $7 + (3/20)$.

This is because the slash, unlike the fraction bar, is not a grouping symbol. It refers only to the numbers immediately preceding and following it. This becomes clearer if the slash is replaced by a fraction bar or division symbol, so that $1/4 - 3$ becomes $\dfrac{1}{4} - 3$ or even $1 \div 4 - 3$. The 3 would only be considered part of the divisor if it were within parentheses, as in the expression $1/(4-3)$, which is equivalent to $\dfrac{1}{4-3}$, which equals 1.

Scientific Notation

Because most standard calculator screens display no more than ten significant digits (and some have room for only six or seven), it is impossible to view very large or very small numbers in decimal form on calculators. For this reason, calculators employ a notation that has long been used in scientific writing: *scientific notation.*

1. Any decimal number can be written in scientific notation as a number between 1 and 10 multiplied by a power of 10. For example:

- $235 = 2.35 \times 10^2$
- $17,800 = 1.78 \times 10^4$
- $0.097 = 9.7 \times 10^{-2}$

The most common powers of 10 are written out in Table 1.12.

2. In scientific notation, the power of 10 indicates how many places the decimal point should be moved to convert the number back to decimal form. Positive powers move the decimal point to the right.

- $2.46 \times 10^2 = 246$
- $3.048 \times 10^6 = 3,048,000$

Negative powers of 10 move the decimal point to the left.

- $1.76 \times 10^{-2} = 0.0176$
- $3.2 \times 10^{-5} = 0.000032$

In general, any number times 10^n stands for that number with the decimal point moved n places: to the right when n is positive, to the left when n is negative.

Table 1.12. Powers of Ten

$10^9 = 10 \times 10 \times 10 \times 10 \times 10 \times 10 \times 10 \times 10 \times 10 = 1,000,000,000$ billions

$10^6 = 10 \times 10 \times 10 \times 10 \times 10 \times 10 = 1,000,000$ millions

$10^5 = 10 \times 10 \times 10 \times 10 \times 10 = 100,000$ hundred thousands

$10^4 = 10 \times 10 \times 10 \times 10 = 10,000$ ten thousands

$10^3 = 10 \times 10 \times 10 = 1,000$ thousands

$10^2 = 10 \times 10 = 100$ hundreds

$10^1 = 10$ tens

$10^0 = 1$ ones

$10^{-1} = \dfrac{1}{10} = 0.1$ tenths

$10^{-2} = \dfrac{1}{10^2} = 0.01$ hundredths

$10^{-3} = \dfrac{1}{10^3} = 0.001$ thousandths

$10^{-4} = \dfrac{1}{10^4} = 0.0001$ ten thousandths

$10^{-5} = \dfrac{1}{10^5} = 0.00001$ hundred thousandths

$10^{-6} = \dfrac{1}{10^6} = 0.000001$ millionths

$10^{-9} = \dfrac{1}{10^9} = 0.000000001$ billionths

3. Because most electronic calculators cannot reproduce standard exponential notation, they employ a variant of it. One such notation employs the letter E (for *exponent*) as follows:

- 2.46×10^2 appears as 2.46E2.
- 1.76×10^{-2} appears as 1.76E-2.

Another notation displays the power of 10 in raised (exponent) position.

- 3.048×10^6 appears as 3.048^{06}.
- 3.2×10^{-5} appears as 3.2^{-05}.

4. Working with Scientific Notation Scientific notation was developed in part for computations involving very large or very small numbers. Calculating in scientific notation requires multiplying and dividing powers of 10. This turns out to involve nothing more than adding and subtracting exponents. To multiply powers of 10, add the exponents:

- $10^9 \times 10^5 = 10^{9+5} = 10^{14}$
- $10^{-3} \times 10^{-7} = 10^{(-3+-7)} = 10^{-10}$

To divide powers of 10, subtract the second exponent from the first:

- $10^6 \div 10^2 = 10^{6-2} = 10^4$
- $10^{-8} \div 10^{-3} = 10^{(-8--3)} = 10^{-5}$

It would be extremely awkward to multiply 28,000,000,000 times 306,000,000 in the standard way. Much simpler would be to find the product of 2.8×10^{10} and 3.06×10^8:

- $(2.8 \times 10^{10}) \times (3.06 \times 10^8) = (2.8 \times 3.06) \times (10^{10} \times 10^8) = 8.568 \times 10^{18}$

In the same way, using standard division to divide 0.0000045 by 0.000025 could be time-consuming. Finding the quotient becomes much simpler in scientific notation:

- $(4.5 \times 10^{-6}) \div (2.5 \times 10^{-5}) = (4.5 \div 2.5) \times (10^{-6} \div 10^{-5}) = 1.8 \times 10^{-1} = 0.18$

5. Significant Digits, Measurement, and Scientific Notation One advantage of scientific notation is that it shows which digits in a figure are significant by eliminating zeros that merely keep place values. Thus 0.108, expressed as 1.08×10^{-2}, shows three significant digits; 480,000 (or 4.8×10^5) shows two.

The number of significant digits depends upon the accuracy of the method of measurement. In many instances a zero may be just as significant as any other digit, as when a pair of calipers graded to the tenth of a millimeter shows a width of 10.00 cm. This is not equivalent to 10 cm, because 10 cm implies an accuracy of 0.5 cm—that is, the measurement is between 9.5 and 10.5 cm. On the other hand, 10.00 cm, which reliably falls between 9.995 and 10.005 cm, implies an error of 0.005 cm.

When zeros are known to be significant in a measurement, scientific notation can emphasize this point. For example, 271.0 kg has four significant digits, whereas 271 kg has three. The first should be written as 2.710×10^2 kg, and the second as 2.71×10^2 kg.

6. Absolute Error and Relative Error   The error in any measurement is one-half of the unit indicated by the last significant digit. A measure of 153 meters is accurate to one meter, which means that there may be an error of 0.5 meters in either direction. This is called the *absolute error.* The absolute error in a measure of 145.7 kg, for example, is ½ of 0.1 kg, or 0.05 kg.

Relative error compares the size of the error to the size of the object being measured. Usually expressed as a percent, it is the ratio of the absolute error in the measurement to the measurement itself.

$$\text{relative error} = \frac{\text{absolute error}}{\text{measurement}}$$

When given as a percent, relative error is referred to as *percent error.*

- If a tree is estimated to be 25 meters tall, give or take 2 meters, the relative error is $\pm\frac{2}{25}$, which is equivalent to a percent error of $\pm 8\%$.

Calculator Calculation

Electronic calculators come in a dazzling array of sizes, shapes, and formats, and a comprehensive guide to their operation is beyond the scope of this book. However, because most standard calculators share features, several useful generalizations can be made. (For more specific instructions, consult the user's manual.)

1. Types of Calculators   Electronic calculators fall into four general types: the basic four-function calculator, the scientific calculator, the graphing calculator, and the specialized calculator, such as a business, engineering, or statistical calculator.

On a basic *four-function calculator* the operations are limited to the four algebraic operations ($+$, $-$, $\times$, and $\div$), and perhaps a percent key or a square-root key. Such a calculator is adequate for only the most rudimentary calculations, and can be replaced at very low cost by a good scientific calculator.

A *scientific calculator* can be distinguished from a four-function calculator by the presence of several important keys marked log, 10^x, ln, e^x, sin$x$, cos$x$, and x^y (or y^x). There may be many other keys, but these will always be present.

At five to ten times the cost of a scientific calculator, a *graphing calculator* represents the high end of the spectrum. While such a calculator can be useful for students and some specialists, it is harder to master than simpler calculators. Its greatest advantage to the average consumer is that it allows formulas and calculations to be entered and displayed on the screen in much the same way they would be written out on paper. Thus it can display an entire calculation, not just its individual elements as on smaller calculators.

A *business calculator* is the standard tool of investment analysts and mortgage underwriters. There are many name brands, each with a different keypad, but all of them include simple routines for calculating mortgage payments, future value of investments, and maturity dates. These procedures are described in Chapter 6.

2. Order of Operations in Calculator Computations Most electronic calculators have the correct order of operations programmed into them. The exceptions are older models that require each operation to be executed in sequence; it is up to the user of an older calculator to type in the operations in the correct order.

- The key sequence $\boxed{5}$ $\boxed{-}$ $\boxed{3}$ $\boxed{\times}$ $\boxed{15}$ $\boxed{\div}$ $\boxed{6}$ $\boxed{=}$ gives a result of 5 if the calculator works from left to right. This will occur if the calculator carries out each operation as it is typed in. If, however, the calculator follows the standard order of operations, it would give a result of -2.5.

In general, it is safest to carry out operations one at a time by pressing = after each operation to assure your control over the order of operations.

3. Operator Keys Several keys carry out operations on numbers displayed on the screen. The simplest examples are the squaring key $\boxed{x^2}$, the square-root key $\boxed{\sqrt{x}}$, and the logarithm key $\boxed{\log}$. Depending on the calculator, these keys should be pressed either before or after a given number is entered.

- The sequence $\boxed{16}$ $\boxed{\sqrt{x}}$ will give the principal square root of 16, which is 4.
- The sequence $\boxed{100}$ $\boxed{\log}$ will give the log of 100, which is 2.

4. The Additive Inverse Key $+/-$ This key is used to change any number that appears in the display window into its additive inverse by changing its sign from positive to negative or negative to positive.

5. The Multiplicative Inverse or Reciprocal Key This key displays the symbol $\boxed{1/x}$, or $\boxed{x^{-1}}$. It changes any number except 0 into its multiplicative inverse. Because the result is given in decimal form, it may not be easily recognizable as a reciprocal. For example, the key sequence $\boxed{\cdot}$ $\boxed{4}$ $\boxed{1/x}$ returns the value 2.5. This is because 0.4 equals $\frac{4}{10}$ or $\frac{2}{5}$, and the reciprocal of $\frac{2}{5}$ is $\frac{5}{2}$, which in decimal form is 2.5.

6. Exponents and Roots To calculate compound interest, population growth estimates, or a geometric mean (see Chapter 5), it is essential to know how to raise a number to an integer or fractional power. This can be done in many ways with a scientific calculator. There are special keys for square roots and cube roots and for squares and cubes. But there is a more versatile pair of keys that can be used to find any other powers or roots. The first is labeled either as $\boxed{x^y}$ or $\boxed{y^x}$, and its counterpart is the key labeled $\boxed{x^{1/y}}$ or $\boxed{y^{1/x}}$ (or possibly $\boxed{\sqrt[x]{y}}$ or $\boxed{\sqrt[y]{x}}$).

To raise one number to the power of another, enter the first number, press $\boxed{x^y}$, then enter the exponent. Some calculators require that the equal sign be pressed; others do the calculation without it.

- To compute 5^4, press the sequence $\boxed{5}$ $\boxed{x^y}$ $\boxed{4}$. The answer is 625.
- To compute 2^{10}, press the sequence $\boxed{2}$ $\boxed{x^y}$ $\boxed{10}$. The answer is 1024.

To find a square root or cube root, use the keys labeled $\sqrt{}$ or $\sqrt[3]{}$, respectively. For a 4th, 5th, or higher root, use the key labeled $\boxed{x^{1/y}}$.

- To compute the 5th root of 24, enter $\boxed{24}$ $\boxed{x^{1/y}}$ $\boxed{5}$. This key sequence represents $24^{1/5}$ or $\sqrt[5]{24}$, and should give the answer 1.888175023. This value is not exact, but it is close enough that, if raised to the 5th power by pressing the sequence $\boxed{x^y}$ $\boxed{5}$, the answer displayed on most calculators will be exactly 24.

To raise a number to a fractional exponent, both the $\boxed{x^y}$ and the $\boxed{x^{1/y}}$ keys may be needed.

- The expression $8^{2/3}$ indicates either the square of the cube root of 8, or the cube root of the square of 8. The order is not important. To carry out the calculation on the calculator, press the sequence $\boxed{8}$ $\boxed{x^{1/y}}$ $\boxed{3}$ $\boxed{=}$ followed by $\boxed{x^y}$ $\boxed{2}$ $\boxed{=}$ to get the answer 4. (As mentioned, the equal signs may or may not be necessary, depending on the calculator.)

7. Degree vs. Radian Measure of Angles Scientific calculators can work with angle measures in *degrees* (in which 360 degrees is a full revolution), *radians* (in which 2π radians represents a full revolution), or *grades* (in which 100 grades is a right angle), according to the user's preference. The default setting is for degrees. (See Chapter 2 for a description of these angle measures as used in trigonometry.)

8. Digital Representation of Numbers In decimal mode, numbers that can be typed into a calculator are limited to 6 to 10 digits (depending on the calculator). When a calculator completes a computation, it will resort to scientific notation for numbers that go beyond this limit.

♦ $\boxed{1}$ $\boxed{\div}$ $\boxed{500}$ $\boxed{=}$ will result in a display of 2^{-03} or 2E-3. A simple trick that will convert this number to decimal form is to add 1, resulting in a display of 1.002. Thus $1 \div 500$ is 0.002.

2 Algebra, Geometry, and Trigonometry

Basic Algebra

Algebra goes beyond arithmetic by using letters in place of numbers to generalize the processes of solving numerical problems. In algebra, questions that can be stated in plain English are instead expressed in mathematical symbols. A question about amount or size, for example, may be translated into algebra's most basic tool—the equation, which is essentially a sentence proposing that two quantities are equal and at least one of the quantities is unknown. The goal of most algebraic procedures is to find solutions to such questions by finding the numerical value of the unknown quantity.

EQUATION SOLVING

1. Algebra allows physical situations to be translated into abstract symbols, stated as equations, and solved. To *solve* means to find the solution set of an equation. A *solution set* consists of all values or numbers which, when substituted for the unknown quantity in any equation, make the equation a true statement.

2. To *solve for* a variable in an equation means to isolate it on one side of the equal sign. This is accomplished using the rules of equation solving given on pages 64-65. To get an idea of the sense of the term "solve for," consider the two equations: $y = x - 1$ and $x = y + 1$. Both describe the same relationship between two numbers designated x and y. That is, they both indicate that the number x is one unit greater than the number y. The first equation is "solved for y," because y stands by itself to one side of the equal sign, and the second is "solved for x," because x stands alone.

Φ ——————————————————————— Φ

THE FATHER OF ALGEBRA

During the reign of Caliph al-Mamun (809–833), Baghdad emerged as the new Athens or Alexandria—a center of scholarship that rivaled any in the Western world. Among the many scientists drawn to al-Mamun's "House of Wisdom" was the mathematician and astronomer Muhammad ibn Musa al-Khwarizmi, whose name and principal work hover over every math student to this day. From the name al-Khwarizmi we get the words *algorism*, which originally referred to the Hindu-Arabic number scheme, and *algorithm*, the term for a numerical procedure or operation such as the method for finding the lowest common denominator. More important, from the title of al-Khwarizmi's greatest work, *Al-jabr wa'l muqabalah*, we got the word *algebra* and the subject itself.

Algebra was not invented by al-Khwarizmi. Just as the Greek mathematician Euclid had synthesized all that was known of geometry, adding many original insights, so al-Khwarizmi borrowed from his predecessors, particularly the Hindus and the Greeks, to produce a comprehensive and influential work. The *Al-jabr* is principally devoted to solving various equations through a method of "restoration," "completion," "reduction," or "balancing" (all of which have been proposed as possible translations of the words in the title). This "balancing" refers to transposing the numbers on either side of the equal sign—adding the same quantity to both sides or subtracting it, multiplying or dividing.

Although he wrote out the names of the numbers instead of using symbols, al-Khwarizmi gave such a detailed description of the Hindu system that he was credited with inventing it. The term *Arabic numerals*, which he inspired, is now gradually being replaced by the more accurate term *Hindu-Arabic numerals*.

Φ ——————————————————————— Φ

3. The equal sign The use of the equal sign is somewhat ambiguous. In the simple arithmetic expression $2 + 3 = 5$, it is used to state a fact. But in $x + 3 = 5$, it is used to state a possibility or even a question, since $x + 3 = 5$ depends on the value chosen for x. "What can x be?" it asks. Algebra provides a way to answer such a question by laying down rules that allow questions (equations) to be transformed into answers (solutions).

Φ ——————————————————————— Φ

FOURSCORE AND SEVEN

In the year 1863 Abraham Lincoln began his most famous speech with the words "Four score and seven years ago." Many people who realize that he was referring to the year 1776 and the signing of the Declaration of Independence might not remember what a *score* is. Fortunately, the sentence can be converted to an equation that leads to the answer.

If a score is designated by the letter x—the unknown quantity—then "four score and seven" means $4x + 7$. This should equal the number of years between 1776 and 1863. This fact is expressed by the equation

$$4x + 7 = 1863 - 1776$$

or

$$4x + 7 = 87$$

This equation has the solution $x = 20$, which means that when 20 is substituted for x, the equation becomes a true statement. Four times twenty plus seven does indeed equal 87, and thus "four score and seven years ago" is just another way of saying "87 years ago," although much more impressively.

Φ ——————————————————————— Φ

When solving equations, one of the best rules any student can follow is: Never place an equal sign between two quantities that are not equal. Although it sounds obvious, it is an easy rule to break.

4. *Terms* are groups of numbers and letters (called *variables*) that are separated by the symbols $+$, $-$, and $=$. In simple equations the variable is contained in one term. In more complex equations several terms might contain the same variable.

- In the equation $3x + 5 = 26$, the terms are $3x$, 5, and 26.
- In the equation $-2y^2 + 7y + 5 = 13$, the terms are $-2y^2$, $7y$, 5, and 13.

5. The numerical part of a term is called a *coefficient*.

- In the term $5x^3$ the coefficient is 5.
- In the term $\frac{4}{3}\pi r^2$ the coefficient is $\frac{4}{3}\pi$.

Remember that coefficients are placed in front of the variable or variables they multiply, and the multiplication symbol is omitted. Thus the product of 7 and the variable x is expressed as $7x$ rather than $x \cdot 7$, and the area of a circle is expressed as πr^2 instead of $r^2 \cdot \pi$.

6. Many equations include one or more terms that contain the unknown or variable. When two or more terms contain the same variable part—that is, the same power of the variable or variables—they are called *like terms*. Because like terms differ only in their coefficients, if the coefficients of like terms are removed, the variable parts that remain will be identical.

- $5x$ and $8x$ are like terms.
- $10x^2y$ and $-4x^2y$ are like terms.
- $4\pi r^3$ and $-2r^3$ are like terms.

7. Like terms can be combined into a single term by combining coefficients.

- $5x + 8x$ becomes $(5 + 8)x$ or $13x$.
- $10y^2 - 4y^2$ becomes $(10 - 4)y^2$ or $6y^2$.

8. Solving an equation requires moving the term or terms that contain the variable to one side of the equal sign, leaving everything else on the other side. Thus the goal of simple equation solving is to *isolate the unknown*. There are four basic rules that may be used in carrying this out.

  i. The same number or quantity may be added to both sides of an equation.
  ii. The same number or quantity may be subtracted from both sides of an equation.
  iii. An equation may be multiplied by the same nonzero number or quantity on both sides.
  iv. An equation may be divided by the same nonzero number or quantity on both sides.

9. Two *equivalent equations* have the same solution set. The process of equation solving is one in which a succession of equivalent equations is derived from the original equation. Any one of the four rules will produce a new equation that is equivalent to the original equation. In the final step of this process the unknown is completely isolated, and the equation is thereby solved.

• To solve the equation $2x + 7 = 15$:
First subtract 7 from both sides: $2x + 7 - 7 = 15 - 7$.
Then simplify: $2x = 8$.
Then divide both sides by 2: $\dfrac{2x}{2} = \dfrac{8}{2}$.
Then simplify to get the solution: $x = 4$.

10. Solving an equation in which the variable appears in several terms requires combining like terms and then isolating the term that contains the variable.

• To solve the equation $5x - 4(x - 4) = 6 + 3x$:
First simplify the left side: $5x - 4x + 16 = 6 + 3x$.
Then combine like terms: $x + 16 = 6 + 3x$.
Subtract x from both sides: $x - x + 16 = 6 + 3x - x$.
Then combine like terms again: $16 = 6 + 2x$.
Subtract 6 from both sides: $10 = 2x$.
Divide both sides by 2: $\dfrac{10}{2} = \dfrac{2x}{2}$.
And simplify to find the solution: $x = 5$.

11. For many students, translating word problems into equations is more difficult than solving the equations themselves. The range of word problem types is almost infinite, but some idea of their difficulty may be conveyed by a rate-time-distance problem.

Suppose a car leaves Toledo heading west at 55 mph. Two hours later another car starts out in pursuit at 65 mph. How long will it take the second car to catch up to the first?

This type of question is fairly typical of rate-time-distance problems. Its solution begins with the recognition that the distances traveled by each car will be the same at the time the second car overtakes the first, yet the second car will have traveled two hours less.

	Rate	×	Time	=	Distance
1st car	55		t		$55t$
2nd car	65		$(t - 2)$		$65(t - 2)$

Set the distances equal to each other and solve for t.

$$65(t - 2) = 55t$$

$$65t - 130 = 55t$$

$$65t - 55t - 130 = 0$$

$$65t - 55t = 0 + 130$$

$$10t = 130$$

$$t = 13 \text{ hours}$$

12. Types of Equations There are several types of equations, and they are usually classified according to the powers of the variables they contain. An equation in which the variable has no exponent is called a *linear equation*. An equation in which the variable is squared is called a *quadratic equation*. An equation in which the variable is cubed is called a *cubic equation*.

- $2x + 3 = 0$ is a linear equation.
- $4x^2 - 8x + 9 = 0$ is a quadratic equation.
- $x^3 + 1 = 0$ is a cubic equation.

Algebra provides solutions to all linear, quadratic, and cubic equations. These are stated as formulas. Each of the formulas results from solving a general equation of each type. This provides a *general solution*, which is stated in terms of the *coefficients* (see page 63). The linear formula is the simplest of the three. The quadratic formula is familiar to any second-year algebra student. The formulas for cubic equations are so complex that they are rarely discussed even in second-year algebra.

The general solutions of linear and quadratic equations are given below along with their solutions, or answers. Note that quadratic equations may have as many as two solutions. The examples that follow show how the formulas work.

linear equation: $ax + b = 0$

solution: $x = -\dfrac{b}{a}$

quadratic equation: $ax^2 + bx + c = 0$

solutions: $x = \dfrac{-b \pm \sqrt{b^2 - 4ac}}{2a}$

• The solution of $3x + 5 = 0$ is given by $x = -b/a$, where $a = 3$ and $b = 5$. The solution is $x = -5/3$.

• The solution of $2x^2 - 13x - 7 = 0$ is given by $x = (-b \pm \sqrt{b^2 - 4ac})/2a$, where $a = 2$, $b = -13$, and $c = -7$. Thus $x = (13 \pm \sqrt{169 + 56})/4 = (13 \pm \sqrt{225})/4 = (13 \pm 15)/4 = -\frac{1}{2}$ or 7.

Coordinate Geometry

Solutions to equations that have only one variable can be plotted on a number line, which is a one-dimensional graph. Equations with two variables lead to graphs that are two-dimensional, three-variable equations have three-dimensional graphs, and so on. Because we live in a world of three spatial dimensions, we have trouble visualizing anything higher-dimensional. In fact we are most comfortable working in two dimensions—that is, on a piece of paper or a computer screen.

1. The most common method of two-dimensional graphing uses a grid on which two perpendicular lines—a *horizontal* and a *vertical axis*—establish a frame of reference. The exact location of any point can be specified by two numbers, or *coordinates*. The first coordinate normally indicates horizontal position (negative numbers indicate *left,* positive numbers indicate *right*), and the second coordinate indicates vertical position (positive means *up,* while negative is *down*). Both use the intersection of the axes, called the *origin*, as their starting point. This is a *rectangular coordinate system.* Coordinate notation is illustrated in the following example.

• The point (2, 5) can be found by starting at the origin—which is identified by the coordinate pair (0, 0)—and moving 2 units to the right and 5 units up. The point (−3, −7) is located by moving from the origin 3 units left and 7 units down.

2. We constantly rely on rectangular coordinates to identify locations. In a city laid out in a grid pattern, addresses may be specified by the nearest intersection of named, numbered, or lettered streets or avenues—the corner of 40th and Spruce streets in Philadelphia, for example, or Fifth Avenue and 52nd Street in New York. Similarly, many games employ grids and coordinates to identify a location by specifying a row and a column; in chess, Q7 is the 7th square in the queen's column. In reading maps, as in the game of Battleship, grid locations are

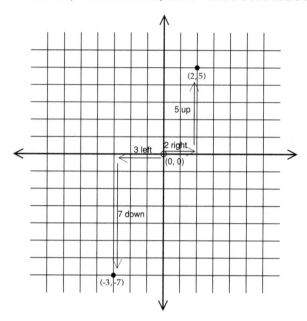

specified by a letter and a number; letters usually run horizontally across the grid, while numbers typically run from top to bottom.

3. The *distance* between any two points on a rectangular grid can be found by using a formula derived from the Pythagorean Theorem (see page 80). Given any two points (a, b) and (c, d):

$$\text{distance} = \sqrt{(a - c)^2 + (b - d)^2}$$

* The distance between the points $(7, -2)$ and $(3, 5)$ is

$$\sqrt{(7 - 3)^2 + (-2 - 5)^2} = \sqrt{4^2 + -7^2} = \sqrt{16 + 49} = \sqrt{65}$$

In three dimensions, each point has three coordinates, and the distance formula is a natural extension of the two-dimensional case. The distance between point (a, b, c) and point (d, e, f) is given by

$$D = \sqrt{(a - d)^2 + (b - e)^2 + (c - f)^2}$$

4. The *midpoint* between two given points (a, b) and (c, d) has the coordinates

$$\left(\frac{a + c}{2}, \frac{b + d}{2} \right)$$

◆ The midpoint between $(2, -7)$ and $(8, 13)$ has coordinates
$$\left(\frac{2 + 8}{2}, \frac{-7 + 13}{2}\right), \text{ or } (5, 3)$$

◆ New York is located at 40°45′N latitude, 74°1′W longitude, and Los Angeles is located at 34°3′N latitude and 118°15′W longitude. Their midpoint can be located by treating latitude and longitude as rectangular coordinates. (See Chapter 4 to review geographical notation.)

latitude of midpoint: $\frac{1}{2}(40°45' + 34°3') = \frac{1}{2}(74°48') = 37°24'N$

longitude of midpoint: $\frac{1}{2}(74°1' + 118°15') = \frac{1}{2}(192°16') = 96°8'W$

These are approximately the coordinates of Kansas City. However, note that lines of latitude and longitude do not truly constitute a rectangular coordinate system because they are drawn on a sphere instead of a plane. Projections of the sphere onto a plane, such as a Mercator projection map (see Chapter 4), do convert latitude and longitude into a grid, but with some distortion. Thus on a Mercator map Kansas City does not appear to be the midpoint of the line that connects New York and Los Angeles, whereas on a globe it does.

5. The steepness of a graphed line is specified through the use of a numerical scale called *slope*. The slope of a line is measured against the horizontal. Specifically, it is the ratio of the vertical rise (or descent) per unit of horizontal travel to the right.

6. A horizontal line has a slope of 0 because it does not rise or fall. A vertical line does not have a slope, since it is impossible to move horizontally along a vertical line. A line that is 45° above horizontal and climbing to the right has a slope of 1, because it rises one unit for each unit it travels to the right. A line that climbs to the left at 45° has a slope of -1, because it falls one unit for each unit it moves to the right.

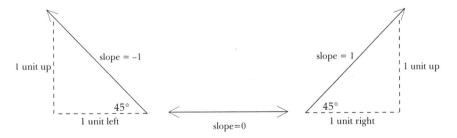

7. Slopes of lines that climb to the right are positive, and slopes of lines that descend to the right are negative.

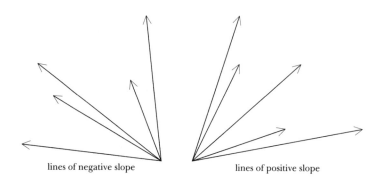

lines of negative slope   lines of positive slope

8. Slope can also be defined as the ratio of the vertical to horizontal displacement between any two points on a line. The vertical displacement is called the *rise,* and the horizontal displacement is called the *run.* This ratio of rise to run is the same for any two points on the line.

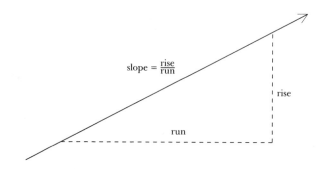

$$\text{slope} = \frac{\text{rise}}{\text{run}}$$

rise

run

9. In construction, engineering, and landscaping, the idea of slope is conveyed by special terms. *Grade,* or *gradient,* refers to the angle of inclination of a surface, a road, or a pipe, expressed as the rise or fall in feet per 100 feet or in meters per 100 meters (both ratios are the same). Grades are usually expressed as percents preceded by a plus or minus sign, indicating rise or fall. A +2% grade is equivalent to a rise of 2 feet every 100 feet. (For more on grade, see page 110.)

   Pitch is a term carpenters use to refer to the steepness of a roof. A roof's pitch is the ratio of rise to run in inches to feet. Thus a pitch of

12 implies a ratio of 12 inches to 1 foot—a 1-to-1 ratio, and thus a 45° angle of inclination. Roof pitches are discussed at greater length in Chapter 3.

Φ ——————————————————————————— Φ

DESIGNING STAIRS

When designing stairs, an architect must keep the tread width and riser height within certain limits in order to satisfy building codes. It is essential that every step be the same, that all risers have the same height and all treads the same width; otherwise, people will trip and fall.

Vitruvius, the first-century B.C. Roman architect, recommended that "The steps in front [of the temple] must be arranged so that there shall always be an odd number of them; for thus the right foot, with which one mounts the first step, will also be the first to reach the level of the temple itself. The rise of the steps should, I think, be limited to not more than ten or less than nine inches; for then the ascent will not be difficult. The treads of the steps are to be made not less than a foot and a half, and not more than two feet deep. If there are to be steps running all around the temple, they should be built of the same size." (*Ten Books on Architecture,* Book III, Chapter IV)

A rise of 9–10″ and a run of 18–24″ make for a stately and leisurely ascent, appropriate for public buildings.

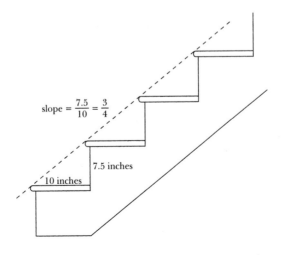

$$\text{slope} = \frac{7.5}{10} = \frac{3}{4}$$

7.5 inches

10 inches

For residential interiors, there are several formulas that permit a narrow range of possible dimensions. These are the restrictions set down by most present-day building codes: (1) two risers + 1 tread = 24–26″, (2) 1 riser + 1 tread = 17–18″, or (3) 1 riser × 1 tread = 70–75″. Taken together, these restrict the riser to 7–8″, and the tread to 10–11″. A common compromise is a rise of 7.5″ and a run of 10″.

Φ ———————————————————————— Φ

EQUATIONS OF LINES

The equation of a line can be found from one of two sets of information: (1) the coordinates of two points that lie on the line, or (2) the slope of the line and the coordinates of a single point. The coordinates of two given points can be plugged into the *slope formula* (see paragraph 1 below), which will reduce the problem to one in which the slope and the coordinates of a point are known. When a point and a slope are known, the equation of the line can be found using *point-slope form*, described below in paragraph 2.

1. If we have the coordinates of two points—(a, b) and (c, d)—the *slope* of the line passing through them is given by the formula

$$m = \frac{d - b}{c - a}$$

Thus the slope (usually designated by m) is the difference in the second coordinates divided by the difference in the first coordinates.

• A line passing through the points $(2, -5)$ and $(6, 3)$ has the slope

$$m = \frac{3 - (-5)}{6 - 2} = \frac{8}{4} = 2$$

The choice of which coordinates to subtract does not matter as long as the coordinates of one point subtract the corresponding coordinates of the other. The slope found above could also be calculated as

$$m = \frac{-5 - 3}{2 - 6} = \frac{-8}{-4} = 2$$

2. The equation in *point-slope form* of a line with slope m that passes through the point (a, b) is given by the formula:

$$y - b = m(x - a)$$

For example, if a line of slope 2 passes through the point (1, 4), its equation in point-slope form would be

$$y - 4 = 2(x - 1)$$

This equation simplifies to

$$y - 4 = 2x - 2$$

Finally, adding 4 to both sides, this becomes

$$y = 2x + 2$$

3. An equation in the form $y = mx + b$ is said to be in *slope-intercept form.* Here the slope is designated by the letter m, and the value b indicates where the line crosses the vertical axis, at a point called the *y-intercept.*

• The equation $y = -3x + 9$ describes a line of slope -3 that crosses the vertical axis at 9.

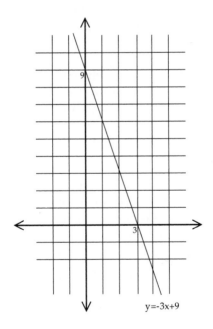

y=-3x+9

4. A *linear relationship* is one in which two quantities are related by an equation such as $y = mx + b$. (It is "linear" because the graph of a straight line expresses the relationship between the quantities.) In the simplest type of linear relationship, one quantity is a multiple of another. (This is also called a *direct proportion.*)

• $y = 1.61x$ is a conversion formula from miles (x) to kilometers (y). Thus 5 miles is about $1.61 \times 5 = 8.05$ kilometers.

In the Appendix, the formula for converting kilometers into miles, or pounds into kilograms, are examples of direct proportions. A more complicated linear relationship is the conversion from degrees Celsius to degrees Fahrenheit; it is expressed by the formula: $F = 1.8C + 32$, which is of the linear form $y = mx + b$. (See Chapter 4 for more on temperature.)

Classical Geometry

Classical geometry, which originated with the ancient Greeks, is the study of the properties of points, lines, and lengths, from which are derived theorems concerning angles, shapes, surfaces, and solids. It was the Greek author Euclid whose treatise *Elements* brought together all of the known geometrical knowledge of his era (c. 300 B.C.). Euclid's text established geometry as a coherent subject, and it remains the basis of the course called *Euclidean geometry* (or *plane geometry*) as it is taught today.

Euclidean geometry is not entirely concerned with drawn figures. It also deals with how a hypothesis leads to a conclusion through a process known as *proof*. That is, although it deals with points and lines, Euclid's geometry is essentially the study of logic—of deductive reasoning. Its theorems rest mostly on basic assumptions called *postulates* and *axioms*. These, along with the definitions of *point, line, plane,* and *length,* combine to form a vast array of conclusions that describe the properties of circles, triangles, squares, rectangles, trapezoids, and three-dimensional solids. These form the basis of the theory of architectural and engineering design. This section is a brief outline of some of the more important geometric definitions and theorems.

DEDUCTIVE REASONING

1. *Deductive reasoning,* which is at the heart of classical geometry, is the basis of *proof,* in which one statement (the *conclusion*) is derived from one or more statements (the *premises*). The process involves a chain of statements connecting the premises to the conclusion in a logical sequence. A simple type of deduction, which Aristotle called a *syllogism,* is a string of connected statements in the form: If A implies B, and B implies C, then A implies C. A classic example of a syllogism is: Socrates is a man; all men are mortal; therefore Socrates is mortal.

Proofs are constructed out of conditional statements. A *conditional statement* is a sentence that puts forth a *hypothesis* in the form of an "if" clause, followed by a *conclusion* or "then" clause. A typical conditional statement is of the form "If *a*, then *b*," which can also be written symbolically as "*a* $\longrightarrow$ *b*" (that is, statement *a* implies statement *b*).

2. Any conditional statement may be true or false, although most statements used in proving other conditional statements either are obviously true or have been established as true. The following are examples of conditional statements whose truth is debatable.

> If a car is green, then it is a Ford.
> If wishes were horses, then beggars would ride.
> If it is raining, then it is not sunny.

3. From any conditional statement of the "If *a*, then *b*" type, three related conditional statements can be constructed from it by reversing the order of hypothesis and conclusion and/or negating both hypothesis and conclusion. These are:

> the *converse:* If *b*, then *a*.
> the *contrapositive:* If not *b*, then not *a*.
> the *inverse:* If not *a*, then not *b*.

4. A conditional statement and its contrapositive are *logically equivalent,* which means that if one is true, then the other must also be true.

> *conditional statement:* If today is Easter, then today is a Sunday. (True)
> *contrapositive:* If today is not a Sunday, then today is not Easter. (True)

5. While the contrapositive of a conditional statement must be true, the converse and inverse are not necessarily true.

> *conditional statement:* If today is Easter, then today is a Sunday. (True)
> *converse:* If today is Sunday, then today is Easter. (False)
> *inverse:* If today is not Easter, then today is not a Sunday. (False)

Because the inverse is the contrapositive of the converse, the inverse and the converse are logically equivalent statements; thus if one is true the other must be true, or (as in the above example) if one is false the other must also be false. The truth of a conditional statement is often established by definitions or postulates—statements whose truth is accepted from the start. The following two sections outline some of the more important definitions and postulates.

DEFINITIONS

A *definition* is a true conditional statement whose converse is also true. The definition "A right angle is an angle measuring 90°" can be put in the form of a conditional statement as "If an angle measures 90°, then it is a right angle." The converse of this would be "If it is a right angle, then it measures 90°." Because the conditional statement and its converse are both true, this constitutes a definition.

In geometry the most basic building blocks—*point, line,* and *plane*—are called *undefined terms* because they are so difficult to state in other words that we have to settle for descriptions such as the following instead of true definitions.

> A *point* is represented as a dot. It is a location in space and it has no dimension.
>
> A *line* can be thought of as a straight thin wire that goes indefinitely in both directions. It has no thickness.
>
> A *plane* is a flat surface that extends infinitely in all directions. It has no thickness.
>
> A *ray* is a line that begins at a point and continues indefinitely in one direction.
>
> An *angle* consists of two rays that share the same endpoint.

The definitions given below, all of which are based on undefined terms, are stated in abbreviated form, not as conditional statements.

Angles

acute angle   an angle whose measure is less than 90°.

right angle   an angle whose measure is exactly 90°.

obtuse angle   an angle whose measure is greater than 90° but less than 180°.

complementary angles   a pair of angles whose measures add up to 90°.

supplementary angles   a pair of angles whose measures add up to 180°.

Circles

circle   the set of points in a plane that are the same distance from a given point. The given point is called the *center,* the distance is called the *radius.*

Triangles

acute triangle   a triangle with three acute angles.

obtuse triangle   a triangle with one obtuse angle.

right triangle   a triangle with one right angle.

scalene triangle   a triangle with no equal sides (or angles).

isosceles triangle   a triangle with at least two equal sides.

equilateral triangle   a triangle with three equal sides.

equiangular triangle   a triangle with three equal angles.

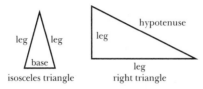

vertex   the point of intersection of two sides. Informally, it is a "corner" of the triangle (although the word *corner* properly refers to a right angle).

altitude   (height) the perpendicular distance from a side to the opposite vertex.

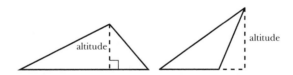

Four-Sided Figures

quadrilateral   any four-sided figure (the sum of its four angles is 360°).

parallelogram   a quadrilateral in which both pairs of opposite sides are equal in length.

rhombus   a quadrilateral with four equal sides (the angles do not have to be equal).

rectangle   a quadrilateral with four right angles.

square   a quadrilateral with four equal sides and four equal angles.

trapezoid   a quadrilateral with exactly one pair of parallel sides.

square   rectangle

rhombus   parallelogram   trapezoid

Regular Polygons

regular polygon   a multisided figure whose sides are of equal length and whose angles are of equal measure.

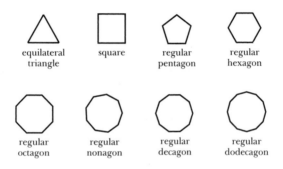

equilateral triangle   square   regular pentagon   regular hexagon

regular octagon   regular nonagon   regular decagon   regular dodecagon

Three-Dimensional Solids While there are an infinite number of three-dimensional solids, there are only five whose faces are all identical regular polygons and all of whose angles are equal. They are called the *Platonic solids,* and are listed below.

tetrahedron   a four-sided figure in which each side (or *face*) is an equilateral triangle.

cube   a six-sided figure also known as a *hexahedron,* in which each side is a square.

octahedron   an eight-sided figure in which each side is an equilateral triangle.

dodecahedron   a twelve-sided figure in which each side is a regular pentagon.

icosahedron   a twenty-sided figure in which each side is an equilateral triangle.

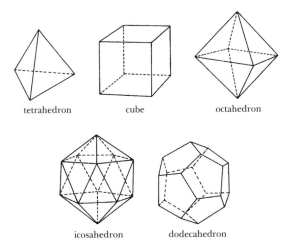

tetrahedron   cube   octahedron

icosahedron   dodecahedron

BASIC POSTULATES

Euclid's *Elements* builds upon five basic postulates (or assumptions), which appear below in paraphrased form.

i. A unique straight line can be drawn between any two points.
ii. Such a line can be extended indefinitely in either direction.
iii. A circle can be drawn in a plane using a given point (a center) and a given distance (radius).
iv. All right angles are equal.
v. Given a line and a point not on the line, there exists exactly one line parallel to the original line passing through the given point.

The fifth postulate, the famous *parallel postulate*, is controversial because, unlike the first four postulates, it is not self-evident, nor can it be proven from the axioms. The acceptance of this postulate leads to what is called *Euclidean geometry;* rejection of it opens doors to many other kinds of geometries, which are referred to as *non-Euclidean geometries*.

The postulates and axioms lead to many conclusions. Below are a few of the more important ones.

Two points uniquely determine a line.

Three points not lying on the same line uniquely determine a plane.

To every pair of points there corresponds a unique positive number referred to as the *distance* between the points.

To every angle there corresponds a unique number called the *measure* of the angle. (Angle measurements may be in *degrees, radians,* or *grades;* see page 108.)

To every region in a plane there corresponds a unique positive number called the *area* of the region.

BASIC THEOREMS

Here are a few examples of statements whose truth is established by the definitions and postulates.

If a triangle is equilateral, then it is isosceles.

If a triangle is not isosceles, then it is not equilateral.

The sum of the measures of the three angles of any triangle is 180°.

The shortest distance from a point to a line is along a perpendicular.

THE PYTHAGOREAN THEOREM

The famous *Pythagorean Theorem* is the basis of the study of right triangles. In words, it says: *The sum of the squares of the lengths of the legs of a right triangle is equal to the square of the length of the hypotenuse.* In symbols:

$$a^2 + b^2 = c^2$$

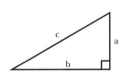

A *Pythagorean triple* consists of three numbers that satisfy the Pythagorean Theorem— $a^2 + b^2 = c^2$. Table 2.1 lists the triples involving numbers up to 50. The starred ones are based on the ratio 3:4:5, which is the most commonly encountered right triangle in textbooks and standardized exams.

Table 2.1. Pythagorean Triples

3, 4, 5*	14, 48, 50
5, 12, 13	15, 20, 25*
6, 8, 10*	15, 36, 39
7, 24, 25	16, 30, 34
8, 15, 17	18, 24, 30*
9, 12, 15*	20, 21, 29
9, 40, 41	21, 28, 35*
10, 24, 26	24, 32, 40*
12, 16, 20*	27, 36, 45*
12, 35, 37	30, 40, 50*

All of the Pythagorean triples can be derived from the following formula, which is based on any two different integer values m and n. The hypotenuse is the third value.

$$(m^2 - n^2); \ (2mn); \ (m^2 + n^2)$$

If $m = 2$ and $n = 1$, then $(m^2 - n^2) = 3$, $(2mn) = 4$, and $(m^2 + n^2) = 5$. The triple 5, 12, 13 results from $m = 3$, $n = 2$.

Area and Volume of Geometric Figures

Size can be measured in many ways. The size of a human being is usually thought of in terms of height and weight, but could also be given in terms of volume, surface area, or even circumference.

For the purpose of measuring, most objects (though not human beings) can be broken down into simple geometric components such as lines, triangles, squares, rectangles, cubes, boxes, circles, spheres, and cylinders. All of these have simple formulas for area and volume, which are outlined below.

1. Perimeter and Circumference The distance around the boundary of any two-dimensional region is called the *perimeter*. The perimeter of a circle is called its *circumference*.

- The perimeter of a triangle or a rectangle is the sum of the lengths of the sides. For an equilateral triangle (in which the sides are of equal lengths), the perimeter is three times the length of a side. The perimeter of a square is four times the length of a side.

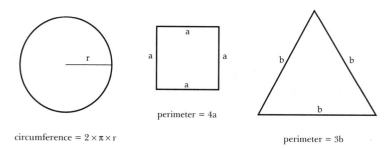

circumference = $2 \times \pi \times r$

perimeter = 4a

perimeter = 3b

- The circumference of a circle is given by the formula: $2 \times \pi \times$ radius, or $2\pi r$. To find the length of an arc of a circle, calculate the ratio of the angle of the arc as a fraction of 360, and multiply this times the circumference, as shown below.

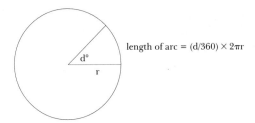

length of arc = (d/360) × 2πr

2. Areas of Squares and Rectangles The area of any rectangle is the product of the length and the width. The area of a square (a special type of rectangle) is the square of the length of a side. Units of area derived from lengths and widths are always given in square units, as shown below.

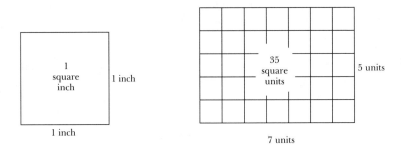

3. Area of a Triangle Any rectangle, when cut in half by a diagonal, forms two right triangles. Because the area of a rectangle is its length times its width (that is, its *base* times its *height*), the area of a triangle formed in this way is calculated as shown below:

area of shaded region
= 1/2 × base × height

This formula holds not just for right triangles, but for all triangles. The units of measure, as in the case of the rectangle, are squared units.

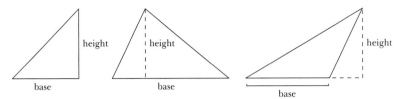

4. Area of a Circle The area of a circle is found by squaring the length of the radius and multiplying the result by π. Thus $A = \pi r^2$. (The approximate value $^{22}/_7$ or 3.14 may often be used for π.)

A *sector* of a circle resembles a slice of a pie. It is formed by an angle drawn from the center of the circle. Its area depends on the measure of the angle; to find the area, take the ratio of the angle measure (in degrees) to 360, and multiply this by the area.

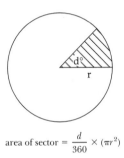

$$\text{area of sector} = \frac{d}{360} \times (\pi r^2)$$

5. Volume of a Solid Any three-dimensional object with vertical sides, a flat top, and a flat base, has a volume given by the formula:

volume = height × area of base

This includes all rectangular boxes (such as the cube), as well as the cylinder. In the case of the cube and other rectangular boxes, the area of the base is equal to the width times the depth. Thus the volume of a box is: *width × depth × height*. Units of volume are always *cubic* units.

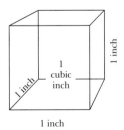

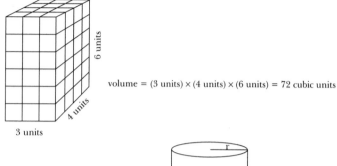

volume = (3 units) × (4 units) × (6 units) = 72 cubic units

volume = $\pi \times r^2 \times h$

6. Surface Area The surface area of a geometric solid can be found by calculating the areas of its sides. For example, a cube, when its sides are laid out flat (as shown below), appears as six sides. A rectangular box also unfolds into six rectangular sides. A cylinder unfolds into two circles and a rectangle.

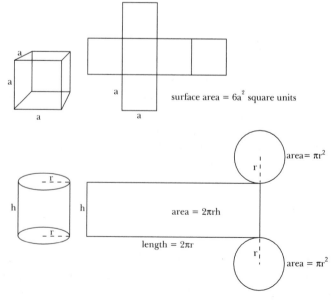

surface area = $6a^2$ square units

area = πr^2

area = $2\pi rh$

length = $2\pi r$

area = πr^2

surface area = $(2\pi r \times h) + (2 \times \pi r)$ square units

Ratio and Proportion ———————————————

RATIOS

1. A *ratio* is an expression of relative size. The ratio of a to b can be written as either $a:b$ or a/b, both of which state a comparison of one object or group of objects to another. A ratio of 2 to 3 can be thought of as 5 units broken into 2 and 3 units.

2. Ratios may be reduced to lowest terms just like fractions. A 10-to-5 ratio is equivalent to a 2-to-1 ratio.

3. The first term of a ratio is called the *antecedent,* the second term the *consequent.* In a 4-to-7 ratio, 4 is the antecedent and 7 is the consequent.

4. The fractional form of a ratio indicates the size of one group in relation to another.

- A pile of 50 poker chips contains 20 red chips and 30 white chips. The ratio of red to white is 20 to 30, or 2 to 3. There are $2/3$ as many red chips as white, and $3/2$ as many white chips as red.

5. The antecedent/consequent form of a ratio indicates how each portion of a ratio relates to the whole. A ratio of a to b implies that a group of $a + b$ objects can be broken into two groups, where a objects fall into one subgroup and b objects fall into the other. The first subgroup is $\dfrac{a}{a + b}$ of the whole, and the second is $\dfrac{b}{a + b}$ of the whole.

- If a social club has 55 members and the ratio of smokers to nonsmokers is 4 to 7, then the group can be divided into 11 groups of 5, where 4 of the groups (or $4/11$) are smokers, and 7 of the groups (or $7/11$) are nonsmokers. $4/11$ of 55 equals 20, and $7/11$ of 55 equals 35. Thus the number of smokers compared to nonsmokers is 20 to 35, which is equivalent to a ratio of 4 to 7.

6. Ratios are sometimes expressed in percent form by converting their fractional equivalents to percents, or by breaking down a quantity into 100 parts.

- A solution composed of 9 cups of water and 1 cup of bleach can be thought of as consisting of 10 parts. The ratio of water in the solution is 9 to 1, which converts to $9/10$, or 90%; the fraction of bleach is $1/10$, or 10%.

♦ A ratio of 3 to 2, when doubled, becomes a ratio of 6 to 4. When multiplied by 10, this becomes the ratio 60 to 40, which converts directly to percent form because its parts add up to 100. That is, the whole can be thought of as consisting of 100 parts which the ratio splits into two groups comprising 60% and 40% of the whole. Ratios based on 100 are referred to by the percent breakdown of the two parts: "sixty-forty," "fifty- fifty," "seventy-thirty," "eighty-twenty," and so on.

7. A triple ratio, written in the form $a:b:c$, states the relative sizes of three quantities. It is composed of two ratios, $a:b$ and $b:c$. The sides of the three squares shown below are in the ratio $1:2:3$. Their areas are in the ratio $1:4:9$.

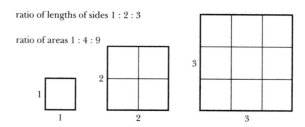

ratio of lengths of sides 1 : 2 : 3

ratio of areas 1 : 4 : 9

PROPORTION

The topic of ratio leads directly to the topic of proportion. Although the two words are often used interchangeably, they have different meanings.

Assume the ratio of vinegar to oil in a salad dressing is 1 to 4. Assume also that you wish to make two cups of dressing. How much oil and vinegar are required?

Problems of this sort come up every day. The idea is to create a mixture or combination in which the ratio of the parts is preserved, so that the oil and vinegar are "in the same proportion"—that is, the amounts of oil and vinegar in the two-cup mixture form a 1-to-4 ratio. This is what is meant by a *proportion*—an equality between ratios.

1. "a is to b as c is to d" is a proportion. It can be written symbolically as $a:b::c:d$, or as $a/b = c/d$. The values b and c are referred to as the *means;* a and d are the *extremes*. In any proportion, *the product of the means is equal to the product of the extremes*. (This is also known as *cross-multiplication*.)

$$\text{If } \frac{a}{b} = \frac{c}{d}, \text{ then } ad = bc.$$

Φ ——————————————————— Φ

THE RULE OF THREE

"Do as you would be done by" is thought to be the one and only Golden Rule. But the name first attached itself to an arithmetic rule for solving proportions, which began to show up in sixteenth-century arithmetic primers. It is known as the Golden Rule to this day in Europe, as well as by its other name, the Rule of Three. It describes the following procedure:

Three quantities are known, of which the first two are in a desired ratio. The goal is to find a fourth term that will make the ratio of third to fourth the same as the ratio of first to second. According to the Golden Rule: Multiply the second and third terms, and then divide by the first. The result is the fourth term. For example, if the three numbers we start with are 2, 3, and 4—that is, if we want to proportionally reproduce the ratio 2:3, with 4 as the first term of the second ratio—then $3 \times 4 = 12$, and $12 \div 2 = 6$. Thus the last number is 6.

A more practical example: If 4 candies cost 70¢, how much do 10 cost? The Golden Rule applied to 4, 70, and 10 gives: $70 \times 10 = 700$, and $700 \div 4 = 175$¢ $= \$1.75$. This, of course, is nothing more than the rule of cross-multiplication stated in words, which makes it a golden rule of thumb.

Φ ——————————————————— Φ

2. When two quantities are *in proportion* or *directly proportional,* it means that one quantity is always the same number times the other. Their ratio is a constant value, and this ratio is called a *constant of proportionality.* In fact, every ratio is a constant of proportionality that may be expressed as a fraction, as a percent, or as a whole-number multiplier. Therefore, to say that quantity A "is proportional to" quantity B is equivalent to saying that A is a multiple of B, or $A = k \times B$. If a train's rate of speed is constant, then the distance it travels is proportional to the elapsed time. This is summed up by the familiar formula: *rate times time equals distance,* which can be written symbolically as $d = r \times t$, where r (the speed) is the constant of proportionality.

◆ When snow melts it turns into water, and the amount (or volume) of water is directly proportional to the depth of the snow. This fact can be used to calculate the weight of a foot of snow per unit of area. The calculation requires two facts: (1) 1 inch of rain weighs 100 tons per acre, and (2) 10 inches of snow is equivalent to 1 inch of rain. Thus a 12-inch snowfall is equivalent to 1.2 inches of rain. This is because the constant of proportionality that converts snowfall to rainfall is $\frac{1}{10}$. The weight of 1.2 inches of rain, if 1 inch equals 100 tons per acre, is 120 tons per acre. Finally, skipping the middle man, the formula that converts inches of powdery snow to tons per acre is $x = 120y$, where x is the weight of snow in tons per acre and y is the depth in feet.

3. If quantity A is *proportional to the square* of quantity B, then their relationship is summed up by the formula: $A = kb^2$. Geometry provides several examples of this type of relationship. Because the area of a square is found by squaring the length of one of its sides, we can say that the area of a square is proportional to the square of the length of a side. In this case the constant of proportionality is equal to 1. The formula for the area of a circle is $A = \pi r^2$. Thus the area of a circle is proportional to the square of its radius, and the number π *(pi)* is the constant of proportionality. (Recall that π is an irrational number approximately equal to 3.14.)

INVERSE PROPORTIONALITY

1. Two quantities are *inversely proportional* (or *inversely related*) when increase in one corresponds to decrease in the other, and their product remains the same. A good example of this is the relation between price and demand for a product. As prices rise, demand falls; as prices fall, demand rises. When this relationship is exact, it may be stated by the equation: $D = k/p$. Thus we say that demand is inversely related to price.

2. If it would take one painter 40 hours to paint a particular house, then it would take two painters 20 hours, four painters 10 hours, five painters 8 hours, and ten painters 4 hours. In each case, the number of painters times the number of hours is the same: 40.

This is an example of inverse proportionality. We would say the time required to paint a house *varies inversely* with the number of painters on the job. Mathematically, this can be expressed as: $x = k/y$, where x is the time required and y is the number of painters, and k is a constant value. In the example above:

$$\text{time required} = \frac{40}{\text{number of painters}}$$

Φ ———————————————————————————— Φ

CONSTANTS OF PROPORTIONALITY
AND CONVERSION FORMULAS

If the speed of light through a vacuum is 186,000 miles per second, how far does light travel in one year?

Whenever the speed of an object is constant, the distance it covers is proportional to the time it takes to cover it. That is: rate (speed) times time equals distance (or $r \times t = d$). In its equivalent form—distance equals rate times time ($d = r \times t$)—this is a proportion where the constant of proportionality is the rate, or speed.

To find the distance light travels in one year, set up the following calculation:

distance = rate × time

= (186,000 miles per second) × (1 year)

Unfortunately, the units of time—seconds and years—do not match. This leaves two choices: convert miles per second to miles per year, or express one year in terms of seconds. We will try the first course.

Converting distance or time units always involves multiplication by a *conversion factor* (which is the same as a constant of proportionality). Feet convert to inches when multiplied by 12. Meters multiplied times 100 convert to centimeters. Since there are 60 seconds in a minute and 60 minutes in an hour, there are 60 × 60 or 3600 seconds in an hour. Therefore,

186,000 miles per *second* = 186,000 × 3600 miles per *hour*

= 186,000 × 3600 × 24 miles per *day*

= 186,000 × 3600 × 24

$\times 365\frac{1}{4}$ miles per *year*

≈ 5.9 billion miles per year

(The symbol ≈ means "is approximately equal to.") Thus light travels a distance of 5.9 billion miles in one year. This distance is known as a *light-year*, which is a convenient unit for measuring the distances to stars and other galaxies.

Φ ———————————————————————————— Φ

3. Many important examples of inverse proportionality come from physics and chemistry. For example, Boyle's Law states that the volume of any constant mass of gas is inversely proportional to the pressure.

$$\text{volume} = \frac{k}{\text{pressure}}$$

If the volume is decreased (or compressed), the pressure increases, and vice versa.

4. Another example of inverse proportionality is the *inverse-square law*, which applies to heat, light, sound, and other forms of radiant energy. To see how it works, consider the candle in the following figure. Its light shines through square openings at A, B, and C; the same amount of light passes through each opening. Window B is twice as far from the candle as window A; thus its edge is twice as long. This means its area is 4 times the area of window A. (See Appendix for area formulas.) Window C, at 3 times the distance, has 9 times the area of A, but the same amount of light reaches it. Therefore the intensity of the light diminishes in proportion to the area, which is related to the square of the distance from the light source.

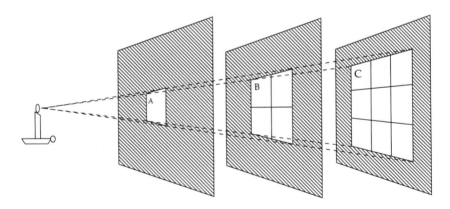

REDIMENSIONING

1. Imagine a rectangular image being projected onto a screen. As the screen is moved further away from the projector, the image enlarges proportionally. The height and width of the image remain in the same proportion, and the same holds true for all lengths within the picture itself. These dimensions will double when the screen-to-projector distance is doubled, triple when the distance triples, and so on.

2. To rescale a rectangle or triangle requires setting up a proportion. One ratio will consist of the height and width of the original; the other will consist of the height and width of the redimensioned image. The problem can be solved if one dimension in the second ratio is known, or if the constant of proportionality (the scale) of the redimension is known.

    ◆ A 12″ × 16″ poster is to be enlarged so that its width is 18″. What is the height of the enlarged poster?

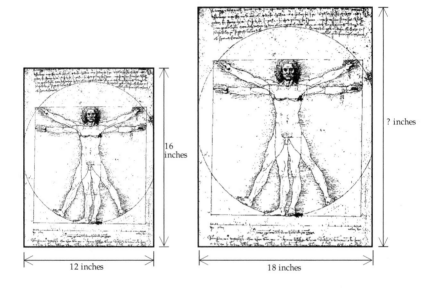

The ratio of height to width for the original poster is 16 to 12. If we divide by 4, this reduces to 4 to 3, or ⁴⁄₃. The ratio of height to width for the enlarged poster is *x* to 18, where *x* is the unknown height. The resulting proportion is

$$\frac{x}{18} = \frac{4}{3}$$

This can be solved by cross-multiplication. (Recall that the product of the means is equal to the product of the extremes.) Thus $3x = 18 \cdot 4 = 72$, and therefore $x = {}^{72}\!/_{3} = 24$.

3. While linear dimensions grow in the same proportion with rescaling, area and volume do not. If the length of a side of a square is doubled, the area of the square is *quadrupled*. If the length of an edge of a cube is doubled, the volume of the cube increases by a factor of 8. (See Appendix for area and volume formulas.)

To see this, start with three lines of lengths 1, 2, and 3:

From these construct three squares:

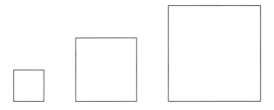

And from these construct three cubes:

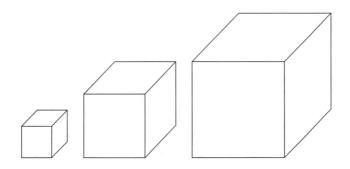

The ratio of the lengths of the sides is 1:2:3.
The ratio of the areas of the squares is 1:4:9.
The ratio of the volumes of the cubes is 1:8:27.

If we want to illustrate the idea of doubling, it is deceptive to draw two squares, one of which is twice the height of the other. The larger square is four times as big, not twice as big. It is even more misleading to draw a cube that is twice the height of another cube, since it would have eight times the volume. This leads to the following guidelines for the interpretation of scaled images:

A visual comparison of one-dimensional objects (lines) is made on the basis of length.

A visual comparison of two-dimensional objects (regions) is made on the basis of area.

A visual comparison of three-dimensional objects (solids) is made on the basis of volume.

Φ ———————————————————————————— Φ

BROBDINGNAGIAN PROPORTIONS

If you increase the height of a man (such as Gulliver) 10 times, so that he stands 60 feet tall, his weight will increase by a factor of 10^3, or 1000. Instead of weighing 10 times his original weight—10 × 200 lbs., or one ton— he would weigh 1000 × 200 lbs., or 100 tons. This thousand-fold increase in weight would be supported on the soles of his feet, which, being a surface, would have increased by a factor of only 100. If Gulliver's foot averages 3″ wide by 1′ long, then in England his two feet had a surface area of 3″ × 12″ × 2, or about 72 square inches. They supported 200 pounds, or about 2.8 pounds per square inch. In Lilliput, the enlarged feet would have to support 28 pounds per square inch (100 tons, or 200,000 pounds, per 7200 square inches), which would make it just about impossible for Gulliver to walk, much less stand up, without breaking his bones.

In anything but a fantasy land, a 60′-tall man would require far different proportions from a 6′-tall man. Jonathan Swift was writing satire, and could overlook the physics implied by his tale. But in the real world, the way creatures look has everything to do with their size and with the laws of proportions.

Φ ———————————————————————————— Φ

4. Two triangles are *similar* if they are identical in shape but not in size. They appear identical in all other respects; the measures of corresponding angles are the same, the ratios of corresponding sides are the same. Thus similar triangles are *proportional* in all linear dimensions.

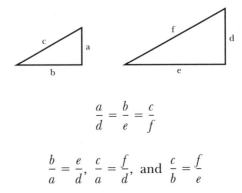

$$\frac{a}{d} = \frac{b}{e} = \frac{c}{f}$$

$$\frac{b}{a} = \frac{e}{d}, \ \frac{c}{a} = \frac{f}{d}, \text{ and } \frac{c}{b} = \frac{f}{e}$$

5. A Simple Measuring Device The proportions of similar triangles can be used to make measurements. A sighting device in the form of a 3-4-5 right triangle, for example, can be used to set up proportions from which the heights of distant objects may be found. The height of the tree in the diagram below can be calculated by pacing off the distance from the base of the tree to the observer. If this distance is 60 feet when the top of the tree is lined up in the sighting device, then the ratio of the tree height to 60 feet must be 3 to 4. The height can be found by solving this equation:

$$\frac{h}{60} = \frac{3}{4} \text{ or } h = \frac{3}{4} \times 60 = 45$$

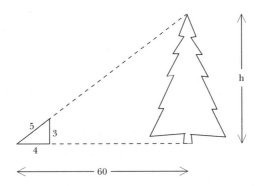

RATIOS AND PROPORTIONS IN EVERYDAY LIFE

1. Gear Ratios When one gear turns another gear, the number of teeth on the driving gear and the number of teeth on the following gear form a ratio called a *gear ratio*. This ratio determines how the rotary force is transmitted from one gear to the next.

A high gear ratio is one in which the driver is larger than the follower. For example, in a gear ratio of 2 to 1, the driver has twice as many teeth as the follower, and twice the circumference. One revolution of the driver will produce two revolutions of the follower.

A low gear ratio involves a smaller driver. In a 1-to-2 ratio, the driver has half as many teeth as the follower, and must make two revolutions in order to produce one revolution in the follower.

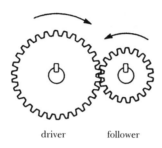

driver   follower

In general, the following ratios hold:

$$\frac{\text{no. of teeth on driver}}{\text{no. of teeth on follower}} = \frac{\text{driver circumference}}{\text{follower circumference}}$$

$$= \frac{\text{no. of revolutions of follower}}{\text{no. of revolutions of driver}} = \frac{\text{follower's speed}}{\text{driver's speed}}$$

In an automobile transmission, gear ratios determine how the rotation of the engine's crankshaft is transmitted to the wheels, which determines how many wheel revolutions are produced by one engine revolution. This is called the *final-drive ratio* and is discussed on page 142.

2. Bicycle Gears Cyclists have their own way of calculating gear ratios that determine a bicycle's performance. A *100-inch gear*, for example, refers to a gear ratio. In this case the front gear has 52 teeth and the rear-wheel gear (or drive gear) has 14 teeth. The ratio of 52 to 14 is multiplied by the wheel diameter, which is 27 inches on racing bikes, to come up with the value 100.

$$\frac{52}{14} \times 27'' = 100.3''$$

A 100.3″ gear is typically referred to as a 100″ gear.

3. Housing Ratios All percents are ratios. Anyone who has applied for a mortgage has heard of the ratio numbers 28% and 36%. The first is the *housing ratio,* which is the accepted ratio of maximum monthly ownership costs to gross monthly income—that is, a maximum threshold that limits the total value of the buyer's monthly mortgage, property tax, and insurance payments to 28% of his gross monthly income. The second ratio is called the *total obligation ratio.* It takes into account *all* monthly debt, which cannot exceed 36% of gross monthly income.

These ratios were established by the Federal National Mortgage Association (Fannie Mae), which regulates how mortgages are resold on the secondary market. Some lenders have relaxed these requirements in recent years. (For more details, see Chapter 5.)

4. SPF Factors Suntanning lotions are graded according to a Sun Protection Factor (or SPF) scale. The scale, which ranges from 2 to 30, represents a multiplicative factor of extra protection from the sun. An SPF of 2 supposedly means that a person who has applied the lotion can remain in the sun twice as long as usual without burning. An SPF of 30 allows the same person to remain in the sun 30 times as long without burning, which means almost indefinitely. Thus a person who can remain in the early-afternoon sun for 15 minutes without burning can remain in the sun for one hour by using a lotion with an SPF of 4. (Scientific evidence of the effectiveness of sun creams is highly suspect. Even though the SPF scale was promulgated by the U.S. Food and Drug Administration, its meaning should be interpreted with caution—it does not provide as much protection as common sense.)

5. *f*-numbers The f-*stop* setting on a camera reduces or enlarges the aperture (or lens opening size), effectively enlarging or reducing the diameter of the lens and thus controls the amount of light admitted when the shutter is opened. The *shutter speed,* by controlling the amount of time the shutter remains open, likewise controls the amount of light to which the film is exposed. The *f*-stop and the shutter speed together define the quality and quantity of light that will create the photographic image.

The *f*-stop (or *f*-number) refers to the ratio of the focal length of the lens (the distance from the lens to the film plane when the camera is focused at infinity) to the diameter of the aperture. The *f*-numbers on a typical camera are $f/1.4$, $f/2$, $f/2.8$, $f/4$, $f/5.6$, $f/8$, $f/11$, $f/16$, $f/22$, $f/32$, and $f/45$. Each setting in this series admits half as much light as the setting that precedes it.

◆ A camera with a focal length of 50 mm is set so that its aperture size is 25 mm. The ratio of focal length to aperture is 50 to 25, which is the same as 2 to 1. The number 2 is referred to as the *f*-number (designated *f*/2). Reducing the aperture from 25 mm to 12.5 mm creates a ratio of 50 to 12.5, or 4 to 1. The *f*-number corresponding to this is *f*/4.

In most cameras the focal length is fixed. Consequently, as the *f*-number is increased, the aperture size decreases. A higher *f*-number implies a higher focal-length-to-aperture ratio, which can only happen if the aperture size is reduced. To maintain the amount of light admitted, the shutter speed can be adjusted to keep the shutter open longer.

In keeping with the inverse-square law (see page 90), the squares of the *f*-numbers are inversely proportional to the amount of light admitted through the aperture. (Thus *f*-numbers are inversely proportional to the radius of the aperture. See Appendix for the formula for circle area.) Table 2.2 shows the *f*-numbers as square roots of index values representing the lens-opening size. Each time the lens-opening size is reduced by half, it generates the next setting.

Φ  Φ

DEPTH OF FIELD

Depth of field refers to a camera's ability to focus on objects that lie in different distance planes. A "high depth of field" permits objects in the foreground and background to be simultaneously in focus. "Low depth of field" creates a sharp focus at a single distance plane; objects that are closer than or farther away from this plane will be out of focus. The depth of the in-focus field increases with the *f*-number. That is, when the aperture size is reduced, the depth of field increases. To achieve this effect, longer exposure times (and thus slower shutter speeds) are necessary. Consequently, to achieve clear pictures with wide depth of field, the aperture should be small (which requires a relatively high *f*-setting) and the shutter speed slow, and the camera must be held very still.

Φ ———————————————————————————— Φ

Table 2.2. *f*-Stop Numbers

Aperture area	f-number
1	1.00
$\frac{1}{2}$	1.41
$\frac{1}{4}$	2.00
$\frac{1}{8}$	2.83
$\frac{1}{16}$	4.00
$\frac{1}{32}$	5.66
$\frac{1}{64}$	8.00
$\frac{1}{128}$	11.31
$\frac{1}{256}$	16.00
$\frac{1}{512}$	22.63
$\frac{1}{1024}$	32.00
$\frac{1}{2048}$	45.25

6. Binocular and Telescope Numbers Binoculars are given two identifying numbers—one that indicates the amount of magnification, and another that gives the diameter of the front lens or lenses in millimeters. The front lens is known as the *objective lens,* or simply as the *objective.* A pair of 6 × 30 binoculars magnifies six times (6×) using an objective lens of 30-mm diameter.

The light-gathering capability, or *relative brightness,* of a pair of binoculars can be found by squaring the ratio of lens size to magnifying power.

$$\text{relative brightness} = \left(\frac{\text{diameter of objective lens}}{\text{magnification}}\right)^2$$

- A 6 × 30 pair of binoculars has a lens-to-magnification ratio of 30 to 6. This can be expressed as the fraction $\frac{30}{6}$, which reduces to 5. The square of this ratio (5^2) gives a relative brightness of 25. A pair of 7 × 25 binoculars has a relative brightness of $(\frac{25}{7})^2$, which is about 13. Thus the second pair of binoculars has about half as much light-gathering capacity as the first.

A simple telescope uses an objective (or front) lens to focus light from a wide beam into a smaller beam that is perceived by the eye. The light-gathering power of the lens is proportional to the area of the objective lens, which involves the square of its radius (and thus the diameter). A dark-adapted eye has an aperture of about $\frac{1}{4}''$ diameter. Thus a telescope with a 2″ objective lens has a diameter 8 times as large as the eye. The area of the objective lens in square inches is $\pi(2)^2$ (see Appendix for circle formulas), and the area of the eye aperture is $\pi(\frac{1}{4})^2$. This ratio is 4π to $\frac{1}{16}\pi$, or 64 to 1. Thus the telescope admits 64 times as much light as the naked eye.

7. Aspect Ratio *Aspect ratio* refers to the dimensions of a rectangle— usually a motion-picture screen, a photograph, a film negative, or some projected image. It is the ratio of width to height. The original standard in the motion-picture industry, established by Edison, was 4 to 3, or about 1.33 to 1. Originally known as the "movietone frame," its adoption by the Academy of Motion Picture Arts and Sciences earned it the name "Academy ratio." In the 1950s, wide-screen moviemaking introduced higher aspect ratios as a response to the threat posed by television (which had inherited its 4 to 3 ratio from Hollywood). Spectacular motion pictures shot in systems such as CinemaScope, Cinerama, and Todd-AO featured ratios varying from 1.65 to 1 up to 2.35 to 1. In 1962, Ultra-Panavision70, an adaptation of CinemaScope technique, provided *Lawrence of Arabia* with an unprecedented 2.75-to-1 aspect ratio. Because the cost of filming such extravaganzas, as well as the problem of outfitting small theaters to show them, proved too great, the industry soon settled on a new standard. Currently, the image ratio in Hollywood is 1.85 to 1; in Europe, it varies from 1.66 to 1 up to 1.75 to 1.

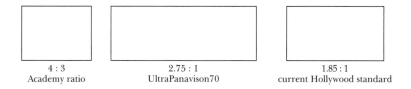

4 : 3
Academy ratio

2.75 : 1
UltraPanavison70

1.85 : 1
current Hollywood standard

8. The Golden Ratio, Golden Section, and Golden Rectangle A ratio of particular importance in classical architecture is the golden ratio. If a line segment is divided into two pieces at a point that makes the ratio of the whole length to the longer piece the same as the ratio of the longer piece to the shorter piece, the division is called a *golden section* and the ratio thus created is called the *golden ratio*.

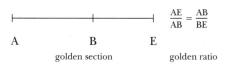

$$\frac{AE}{AB} = \frac{AB}{BE}$$

A   B   E

golden section   golden ratio

Φ ———————————————————————————————————— Φ

EXTRA POSTAGE

Greeting-card designers will try anything to stand out from the pack. But when they succeed and you buy one of their cards, it might very well come back to you because of insufficient postage. Just because it's a greeting card does not automatically mean that it will cost no more to send than an ordinary letter. What the Postal Service means by "ordinary" is quite precise and well worth keeping in mind.

A "standard size" envelope must be rectangular, with a height between 3¼" and 6⅛", and a width between 5" and 11½". But that's not all. What most people fail to realize is that there is also a restriction on aspect ratio. Extra postage is required for any envelope whose length-to-height ratio is less than 1.3 to 1, or greater than 2.5 to 1. In other words, beware of any cards that are too long or too square.

Φ ———————————————————————————————————— Φ

The numerical value of the golden ratio is exactly ½(1 + √5), which is about 1.618 to 1. This is easier to picture as a ratio of about 8 to 5.

A *golden rectangle* is a rectangle whose aspect ratio is a golden ratio. That is, the exact ratio of width to height is ½(√5 + 1) to 1. Such a rectangle represents the ideal of proportion in classical design. Its dimensions are considered the most aesthetically pleasing of all rectangles, and it was widely used by the ancient Greeks in designing temples such as the Parthenon.

To construct a golden rectangle, begin with a square such as square ABCD below. Find the midpoint of AB and call it M. Draw the diagonal from M to corner C. Use this diagonal as the radius of an arc centered at M and drawn down to intersect the extension of line AB. Label the

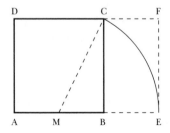

intersection E. Extend a perpendicular from AE to where it intersects the extension of DC at point F.

In the diagram, ᴬᴱ/ᴇғ is a golden ratio, as is ᴬᴱ/ᴀʙ. If the square ABCD is removed, what remains (rectangle CBEF) is also a golden rectangle.

9. Strength of a Beam How much weight will a beam support? It makes sense that the thicker it is, the stronger it is. But by how much? If you want more bearing strength in a beam, for example, do you choose a wider beam or one with more depth? (Depth refers to the dimension that is in the same direction as the force of the load, which is vertical; the width is the dimension parallel to the force.)

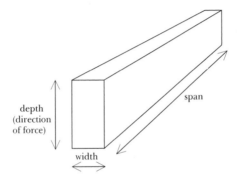

As you may have guessed, depth is more important than width, and this is borne out by proportions. The strength of a beam is (1) directly proportional to its width, (2) directly proportional to the *square* of its depth, and (3) inversely proportional to its length.

Thus,

$$\text{weight capacity} = k \times \frac{\text{width} \times \text{depth}^2}{\text{span}}$$

Consider what this says. If you choose a beam of greater width, the weight it can support increases in proportion to the increase in width. But if you double the depth, for example, you quadruple the bearing capacity (because the increase in strength is proportional to the *square* of the depth). Thus depth is far more critical to the strength of the beam. Notice also that as you increase the span, the capacity *diminishes* proportionally.

What follows is an example of the calculation of the load-carrying capacity of a single beam. The calculation would be different if this were to be one in a series of equally spaced beams. The math involved is very straightforward. If you were building a deck, for instance, you should be able to calculate the joist size yourself by using a calculator and a carpenter's reference guide, which gives the formulas, live load capacities, and stress constants for different types of lumber.

◆ Assume that a No. 1 Douglas fir 2″ × 8″ board must span 10 feet. How much can it hold? The actual width is 1.5″ and the depth is 7.5″ (see page 135 for an explanation of "true" board dimensions). The span is 120″. The value of k, the constant of proportionality, is 2000 for this situation (this figure comes from a table in a carpenter's reference guide). Thus

$$\text{weight capacity} = 2000 \times \frac{(1.5) \times (7.5)^2}{120} \approx 1400 \text{ lbs.}$$

Trigonometry

The word *trigonometry* means "triangle measurement." It refers to the study of the relations between the lengths of the sides of a triangle and the measures of its angles. It is used for calculating heights, distances, and angles as they arise in architecture, engineering, surveying, and navigation, and it is fundamental to quantum mechanics, electromagnetism, and acoustics. It may be viewed as a direct extension of the geometry of familiar triangles such as the 30°-60°-90° and right isosceles triangles.

This section is a brief survey of trigonometry. Its focus is on the analysis of right triangles and their use in measurement and estimation. It defines the trigonometric functions of sine, cosine, and tangent, and explores how electronic calculators can be used to solve triangle measurement problems.

DEFINITIONS AND NOTATION

A *triangle* has three angles, three sides, and three vertices. Capital letters are traditionally used to label each vertex, and lowercase letters denote the lengths of the sides. The capital letters are also used to designate angles.

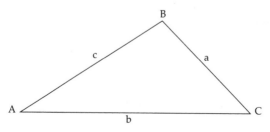

Each angle of a triangle has an *opposite side* and two *adjacent sides*. In the figure above, the side opposite ∠A is BC, which has length *a*. AB and AC are the adjacent sides.

1. Right Triangles A *right triangle* contains one right angle. Because the three angles of a triangle must sum to 180°, the two other angles must sum to 90°. Thus they are both acute angles (less than 90°), and are also complementary angles (they combine to form a right angle).

2. In a right triangle the two sides adjacent to the right angle are called the *legs*. The side opposite the right angle is called the *hypotenuse*. By the Pythagorean Theorem, the sum of the squares of the legs of a right triangle equals the square of the hypotenuse ($a^2 + b^2 = c^2$).

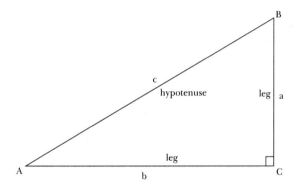

3. A *right isosceles triangle* has angles measuring 45°, 45°, and 90°. The ratio of its sides is $1:1:\sqrt{2}$, as illustrated in the following figure.

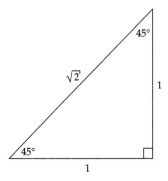

The ratio of the legs is 1 to 1. The ratio of the hypotenuse to a leg is $\sqrt{2}$ to 1.

4. A *30°-60°-90° right triangle* has sides in a ratio of $1:\sqrt{3}:2$. The short leg is half the length of the hypotenuse. The longer leg is $\sqrt{3}$ times the length of the shorter leg.

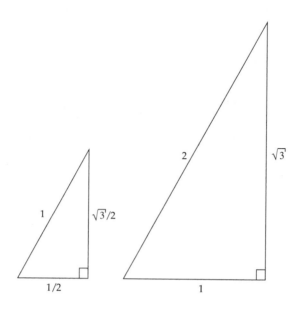

5. An *angle of elevation* measures a line of sight that lies above the horizontal. An *angle of depression* measures a line of sight that lies below the horizontal. If the angle of elevation of the sun is 30°, a tall object such as a tree, flagpole, or building will cast a shadow that forms a 30°-60°-90° triangle.

• Imagine that a flagpole casts a 90′ shadow when the sun is at a 30° angle of elevation. The height of the flagpole can be found by using the proportions of a 30°-60°-90° triangle. The height of the pole is to the length of the shadow as 1 is to $\sqrt{3}$.

$$\frac{x}{90'} = \frac{1}{\sqrt{3}}$$

Multiply both sides by 90′ to get:

$$x = \frac{1}{\sqrt{3}} \times 90' \approx 51.96, \text{ or about } 52'.$$

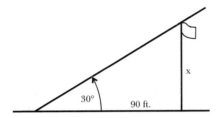

TRIGONOMETRIC FUNCTIONS

The trigonometric functions are used to find the lengths of sides of triangles and triangular regions, or the distance between objects that form the corners of a triangle. If the length of just one side of such a triangle is known, and if the angles are also known, the lengths of the other sides can be determined by using the values of these "triangle measurement" functions.

The trigonometric functions—sine, cosine, tangent, cosecant, secant, and cotangent—are defined as ratios that can be constructed from the sides of any right triangle. Because there are three sides, there are six possible ratios that can be constructed:

> side 1 to side 2
> side 1 to side 3
> side 2 to side 3
> side 2 to side 1
> side 3 to side 1
> side 3 to side 2

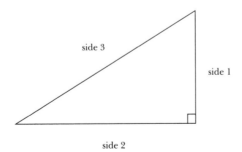

The values of these ratios depend on the measures of the angles; so if the length of one side is known, the lengths of the other sides can be found from the angles using the values of the trigonometric functions. See "Solving Right Triangles" below.

Three of the ratios are *reciprocals* of the other three (see page 34 for an explanation of reciprocals). For example, *b* to *a* (which can be expressed as b/a) is the reciprocal of *a* to *b* (a/b). To simplify the discussion, only three of the trigonometric ratios are addressed here—the sine, the cosine, and the tangent.

Consider the series of superimposed right triangles in the following figure. In each triangle, the ratio of height to base is related to the angle opposite the height, and is called the *tangent* of the angle. If the height increases while the base remains the same, the tangent ratio increases.

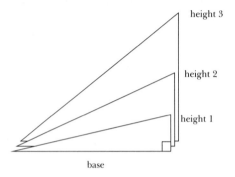

height 3

height 2

height 1

base

The *tangent* (abbreviated *tan*) of an acute angle of a right triangle is the ratio of the opposite side to the adjacent side.

$$\tan\theta = \frac{a}{b}$$

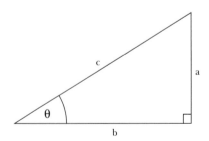

c

a

θ

b

Two other ratios depend for their values on the size of the acute angle at the base. The *sine (sin)* of an acute angle of a right triangle is the ratio of the opposite side to the hypotenuse.

$$\sin\theta = \frac{a}{c}$$

The *cosine (cos)* of an acute angle of a right triangle is the ratio of the adjacent side to the hypotenuse.

$$\cos\theta = \frac{b}{c}$$

1. Solving Right Triangles The sine, cosine, and tangent of an angle can be used to find the dimensions of a right triangle when only one of the sides is known. Table 2.3 shows these values for some common angles. (For an extended version of this table, see the Appendix.)

Table 2.3. Trigonometric Function Values

angle θ	sin θ	cos θ	tan θ
0°	0	1	0
15°	.2588	.9659	.2679
30°	0.5	.8660	.5774
45°	.7071	.7071	1.000
60°	.8660	0.5	1.7321
75°	.9659	.2588	3.7321
90°	1	0	—

The table reveals some interesting relationships. Recall that two complementary angles add up to a right angle (90°); thus the pairs 15° and 75°, 30° and 60°, and 45° and 45° are complementary pairs. Notice also that the sine of any one of these angles is equal to the cosine of its complement; thus the sine of 30° is equal to the cosine of 60°. This is true not just for these pairs but for any complementary pair, and it leads to a general rule: *The sine of any acute angle is equal to the cosine of its complement, and vice versa.* Thus the "co-" in *cosine* means essentially "complement."

2. The values of sine, cosine, and tangent may be used to find the lengths of sides of right triangles in the following ways:

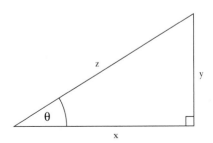

If the hypotenuse z is known, the cosine of angle θ can be used to find x with the formula

$$x = z \cdot \cos\theta$$

If the length x is known, the tangent of angle θ can be used to find y with the formula

$$y = x \cdot \tan\theta$$

If the length y is known, the sine of angle θ can be used to find z with the formula

$$z = y / \sin\theta$$

ANGLE MEASUREMENT

1. Degree Measure Angles may be measured in any of three ways. The most common is *degree measure,* which originated in antiquity. The others are *radian measure,* used primarily by mathematicians, and *grade,* a metric angle measure used almost exclusively by engineers.

A *degree* is an angle measurement equivalent to $\frac{1}{360}$ of a full revolution.

A *minute* is $\frac{1}{60}$ of a degree.

A *second* is $\frac{1}{60}$ of a minute, and thus $\frac{1}{3600}$ of a degree.

32° 14′ 32″ means 32 degrees, 14 minutes, 32 seconds. When an angle is so small that its measure is given in minutes or seconds alone, they are generally written as *minutes of arc* or *seconds of arc* to avoid any confusion with units of time. (See section beginning on page 157.)

2. Radian Measure *Radian measure* is an alternative to degree measure. It equates a full revolution with the measure 2π. Thus 2π radians is equivalent to 360°. (See Appendix for the common circle formulas.) Its usefulness lies in its relation to the length of the arc of a circle intercepted by the angle. In a circle of radius 1, the radian measure of a central angle (an angle with its vertex at the center of the circle) is the same as the length of the intercepted arc. In general, the arc length intercepted by a central angle of any circle is equal to the radius times the radian measure of the angle. This is illustrated in the following figure.

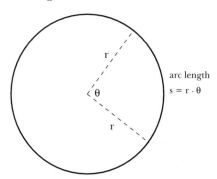

arc length

$s = r \cdot \theta$

Φ ———————————————————————— Φ

BEARING AND RANGE

To "find your bearings" means to figure out where you are and in what direction you are headed. The phrase comes from a more precise meaning of *bearing,* which comes up most often in the context of war. An artillery-man seeks to locate a target using two coordinates: direction and distance—that is, bearing and range.

A *bearing* is an angle that specifies direction with respect to a fixed direction such as the north-south axis. A bearing angle is never greater than 90°, and thus it must specify a *quadrant.* From where you are sitting, there are two imaginary lines, one running north to south, the other running east to west, which establish four quadrants: the northeast, southeast, southwest, and northwest quadrants. A bearing of N60°E clearly specifies the northeast quadrant. The angle is measured clockwise from due north. The figure below shows four bearing angles and how they are measured either from the north or south, depending on the quadrant.

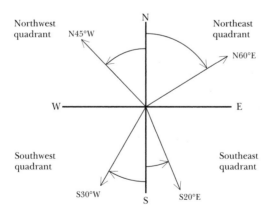

The *range* is a distance along a particular bearing. For example, a target at a range of 2000 meters and a bearing of N75°E can be located by aiming 75° clockwise from due north and setting the distance at 2000 meters. This can be translated into rectangular coordinates using sines and cosines. In other words, the object could be reached either by setting off as the crow (or artillery shell) flies or (assuming you are in a city laid out in a grid pattern, such as New York, and have to take a cab) traveling east a certain distance and then north a certain distance. These distances—the rectangular coordinates—are given by the following formulas:

No. of meters east: $2000 \times \sin 75°$

No. of meters north: $2000 \times \cos 75°$

That is, multiply the sine of the bearing angle by the range to get the east-west distance, and multiply the cosine of the bearing by the range to get the north-south distance.

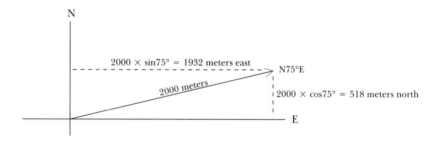

For more on bearings and the related topic of azimuths, see page 174.

Φ ———————————————————————— Φ

3. Grade A *grade* is equal to $\frac{1}{100}$ of a right angle, or $\frac{1}{400}$ of a full revolution. One of the original metric units of measure, the grade is rarely encountered today outside of engineering contexts, where it is expressed as a percent (and sometimes called the *gradient*).

♦ In rapid-transit rail systems, the maximum grade allowed between two stations is 3.5%. The maximum allowable grade *through* a station is 0.3%. This means that the tracks should not rise or fall more than 3.5 meters per 100 meters between stations. And, assuming a station is 100 meters long, the tracks should lose no more than 0.3 meters between one end of a platform and the other.

4. The conversion formulas for degrees, radians, and grades are:

$$1 \text{ degree} = \frac{\pi}{180} \text{ radians} = \frac{10}{9} \text{ grade}$$

$$1 \text{ radian} = \frac{180}{\pi} \text{ degrees} = \frac{200}{\pi} \text{ grade}$$

$$1 \text{ grade} = \frac{\pi}{200} \text{ radians} = \frac{9}{10} \text{ degree}$$

5. Most electronic calculators allow angle measurements to be entered either in degree-minute-second format or in radian measure. When an angle is entered using the degree-minute-second key (° ′ ″), it is usually converted to a decimal form. If your calculator does not have such a key, angle measures in degrees-minutes-seconds may be converted to decimal form by using fractions.

- 30° 12′ is equivalent to 30 and $^{12}/_{60}$ degrees. Since $^{12}/_{60}$ equals $^{1}/_{5}$, or 0.2, the angle measure of 30°12′ will be displayed by a calculator as 30.2°.
- 56°6′36″ is equivalent to $56 + {}^6/_{60} + {}^{36}/_{3600}$, or $56 + 0.1 + 0.01$, or 56.11°.

Scientific calculators can easily switch back and forth between degree, radian, and grade modes. The screen will indicate the mode by showing the abbreviations DEG, RAD, or GRAD. It is important to note the mode when calculating values of trigonometric functions.

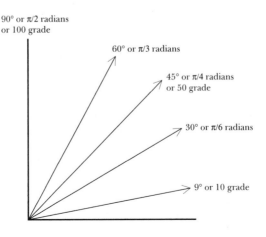

Table 2.4. Degree, Radian, and Grade Equivalents

0°	0 radians	0 grade
15°	$\pi/12$ rad	16.67 grade
30°	$\pi/6$ rad	33.33 grade
45°	$\pi/4$ rad	50 grade
60°	$\pi/3$ rad	66.66 grade
75°	$\pi/12$ rad	83.3 grade
90°	$\pi/2$ rad	100 grade

A Glance at Calculus

These first two chapters have concentrated on the basic principles of arithmetic, geometry, algebra, and trigonometry. Properly fleshed out, these topics constitute a mathematical education through the high-school level, and for some students even through the college level. Traditionally, college-level mathematics begins (and, in many cases, ends) with calculus, but most students never take it, and few go beyond it.

A common assumption about calculus is that it is self-contained. Few students understand the extent to which calculus is a direct continuation of algebra. Some idea of this connection can be conveyed by a simple example.

Imagine a car traveling down a highway from point A to point B. The car has a device (perhaps an odometer hooked up to a clock) that records its exact position at every instant of time. The driver stops occasionally to refuel or to eat, and this is what makes the problem different from a typical algebra word problem: the speed of the car is not constant. Thus the formula "rate times time equals distance" doesn't apply here because the rate—the speed of the car—keeps changing. The question is: Can we compute the car's speed from its position? We know that if a car covers 50 miles in one hour its *average* speed is 50 miles per hour. But what if the car did not maintain the same speed through the entire trip? Can we determine its speed at 5, 10, or 20 minutes into the trip? This goes far beyond what algebra alone can handle, but calculus provides a technique to solve it.

Calculus has been defined as the mathematics of change. In fact, as soon as the concept of "change" is introduced into algebra, it becomes calculus. Rates of change (speeds, rates of disease transmission, population growth) are the subject of what is called *differential calculus.* Its counterpart, *integral calculus,* can be appreciated using a slightly different view of the problem outlined above.

Assume that a car records speedometer readings (instead of its position) at every point during a trip. Using only this information, can the distance the car has traveled be calculated? In other words, can we compute its position from its speed? This question is easy if the speed remains constant: according to algebra, distance equals rate times time. But what if the speed keeps changing? This presents a more difficult problem, beyond the realm of algebra but well within the grasp of calculus.

Calculus maps out a vast territory packed with applications and implications that these examples only begin to suggest. As illustrated in the story of Sir Isaac Newton (its inventor) and the apple tree, the purpose of calculus is to explain how change takes place—how fast the apple falls, how long it takes to fall. Such questions led thinkers like Newton to invent techniques that allow rates of change and amounts of change to be measured. This led to a scientific revolution, which helps to explain why calculus is the basic tool of almost all scientific investigation, and a required course for any college student who wishes to pursue research in the physical sciences or life sciences.

The process of mathematical discovery continues to unfold to this day. New areas of investigation, such as fractals and chaos, yield further clues to the nature of the world we inhabit. Such topics may be mathematically complex (many people might describe even basic algebra as complex), and perhaps can be appreciated by nonexperts only through the graphic images they produce. But they do contribute to everyday conversation and discussion. When we talk about the world around us— about science—the language we use is the language of science, which is mathematics.

3   Everyday Mathematics

The present chapter is titled "Everyday Mathematics" because it deals with four familiar areas of everyday life in which math questions regularly arise. But as the book's title suggests, much of the rest of the text is likewise devoted to practical, everyday math, and numerous discussions of such problems are scattered throughout. The reader looking for help with a given problem should refer to the Table of Applications on the inside front cover as well as to the index.

Food and Drink

Food shoppers must navigate a sea of numbers. Package sizes, weights, volumes, nutritional values, fat content, special discounts, expiration dates, recipe requirements, taxes, and food budgets are just some of the considerations that arise in a trip to the market or supermarket. Understanding these numbers requires some familiarity with ratios, proportions, percents, decimals, markups, discounts, sums, and number comparisons. Not all of these topics are addressed in this section. Decimals, percents, fractions, markups, and discounts are discussed in Chapter 1. Ratios and proportions are covered in Chapter 2. This section concentrates on packaging—on size or amount, nutritional content, and unit prices. For some shoppers these are crucial considerations, for others merely a distraction. Most would rather let the cash register handle the accounting and the palate make most of the choices. The government, at least, would like everyone to have access to a few key numbers, which is why it requires certain nutrition facts to be published on each package of food. These numbers are discussed below.

NUTRITION LABELING
New Food and Drug Administration guidelines issued in 1994 put forth two sets of dietary recommendations—one a set of *maximum daily values*

of food groups, the other of *minimum daily values* of vitamins and minerals. ("Daily Value" has superseded the older terms "Minimum Daily Requirement" and "Recommended Daily Allowance," since it is a vaguer and more neutral term that seems appropriate for describing both maximum and minimum desirable amounts.) The guidelines also mandated that every packaged food be labeled with a statement of *nutrition facts,* listing maximum and minimum daily values of vitamins, minerals and other nutrients. Although the statement of Nutrition Facts is supposed to be self-explanatory, a few observations are worth emphasizing.

Tables 3.1 and 3.2 give the FDA's recommendations for nutrients, vitamins, and minerals, all based on a daily diet of 2000 calories. The statement of "Nutrition Facts" that appears on food packaging breaks down the content of one serving and relates it to the tables of Daily Values. These percentages are also based on a 2000-calorie diet.

A typical statement of nutrition facts is shown below.

Nutrition Facts

Serving Size 3 biscuits (34 g)
Servings per Container About 6

Amount Per Serving

Calories 170 Calories from Fat 60

	% Daily Value*
Total Fat 7g	11%
Saturated Fat 3g	15%
Cholesterol 10mg	3%
Sodium 170mg	7%
Total Carbohydrate 24g	8%
Dietary Fiber 1g	4%
Sugars 5g	
Protein 2g	

Vitamin A 0% · Vitamin C 0%

Calcium 0% · Iron 6%

*Percent Daily Values are based on a 2,000 calorie diet. Your daily values may be higher or lower depending on your calorie needs:

	Calories:	2,000	2,500
Total Fat	Less than	65g	80g
Sat Fat	Less than	20g	25g
Cholesterol	Less than	300mg	300mg
Sodium	Less than	2400mg	2400mg
Total Carbohydrate		300mg	375mg
Dietary Fiber		25g	30g

Calories per gram:
Fat 9 · Carbohydrate 4 · Protein 4

Table 3.1. FDA Dietary Recommendations (based on a daily diet of 2000 calories)

Food group	Maximum daily value
Fat	65 g
Saturated fatty acids	20 g
Cholesterol	300 mg
Sodium	2400 mg
Total carbohydrate	300 g
Fiber	25 g
Potassium	3500 mg
Protein	
Infants	14 g
Children 1 to 4 yrs.	16 g
Pregnant women	60 g
Nursing mothers	65 g
Others	50 g

Table 3.2. FDA Minimum Daily Values of Vitamins and Minerals

Vitamin A	5,000 IU (international units)
Vitamin B6	2 mg
Vitamin B12	6 micrograms
Vitamin C	60 mg
Vitamin D	400 IU
Vitamin E	30 IU
Folic acid	0.4 mg
Thiamine	1.5 mg
Niacin	20 mg
Riboflavin	1.7 mg
Calcium	1.0 g
Iron	18 mg
Phosphorus	1 g
Iodine	150 micrograms
Magnesium	400 mg
Zinc	15 mg
Copper	2 mg
Biotin	0.3 mg
Pantothenic acid	7 mg

1. Minimum vs. Maximum Daily Values—How Much Should You Eat?
The FDA makes two very different types of recommendations in its tables. Percentages of fat, cholesterol, sodium, and carbohydrates are *maximum* allowances, while percentages of vitamins and minerals are

minimum allowances. Seven grams of fat, for example, constitutes 11% of the maximum recommended intake of 65 grams (7 grams divided by 65 grams is about 0.1077, which can be rounded to 11%). The 11% in the right-hand column does not mean that this food contains 11% fat.

Daily Values for fat, cholesterol, sodium, carbohydrates, and dietary fiber are itemized on some packages for both a 2000- and a 2500-calorie diet. The 2000-calorie group includes most women and teenage girls, and less-active men. The 2500-calorie level targets active men and teenage boys, as well as very active women. The small print at the bottom of the Nutrition Facts table gives calorie-per-gram equivalents for fat, carbohydrates, and protein.

2. Fat Content The FDA recommends a total fat intake of no more than 30% of all calories consumed. Some nutritionists think this is too strict, but no one places the value above 40%. Keep in mind that: *1 gram of fat converts to 9 calories, whereas 1 gram of protein or carbohydrate yields about 4 calories.* Thus fat has more than twice the per-gram calories of carbohydrate or protein.

The percent fat content is not listed in the Nutrition Facts table. It can easily be calculated, however, from two values listed near the head of the table: calories per serving and calories from fat.

$$\% \text{ fat content} = \frac{\text{calories from fat}}{\text{calories per serving}}$$

♦ A jar of dry-roasted peanuts contains 160 calories per 28-gram serving, of which 14 grams constitutes fat, which is shown as 22% of the maximum Daily Value of fat you should consume. The label also states that this 14 grams of fat amounts to 120 calories. If one serving totals 160 calories, the proportion of the peanuts' calories that comes from fat is $120/160$, or a whopping 75%.

3. Sodium Content The recommended sodium intake per day— between 2400 and 3000 mg—can be quickly met and easily exceeded. One teaspoon of salt contains about 2300 mg of sodium, and most foods naturally contain salt. Soups, for example, tend to be very high in sodium, as do most snacks.

4. Serving Size Be sure to note the serving sizes indicated at the top of the Nutrition Facts table. Compare these to your typical intake and adjust the Daily Values accordingly. If three biscuits contain 7 grams of fat, and you usually eat a dozen, multiply by 4 to find that your typical serving contains 28 grams of fat.

The U. S. Department of Agriculture has established a Food Guide Pyramid, which lists recommended daily servings of each major food group. From bottom to top, the pyramid sets these goals:

Bread, cereal, rice, and pasta group	6–11 servings
Vegetable group	3–5 servings
Fruit group	2–4 servings
Milk, yogurt, and cheese group	2–3 servings
Meat, poultry, fish, beans, eggs, and nuts group	2–3 servings
Fats, oils, and sweets group	Use sparingly

Serving sizes are given as follows:

Bread	1 slice
Dry cereal	1 ounce
Cooked cereal, rice, pasta	½ cup
Leafy vegetables	1 cup
Other vegetables (cooked or uncooked)	½ cup
Vegetable juice	¾ cup
Fruit	1 apple, banana, or orange
Chopped, cooked, or canned fruit	½ cup
Fruit juice	¾ cup
Milk or yogurt	1 cup
Natural cheese	1.5 ounces
Processed cheese	2 ounces
Cooked lean meat, poultry, or fish	2–3 ounces

(½ cup cooked dry beans, 1 egg, 2 tablespoons of peanut butter, or ⅓ cup nuts are each equivalent to 1 ounce of meat.)

ALCOHOL

1. Alcoholic Proof The alcohol content of distilled liquor is measured by *proof*—a number between 0 and 200 representing twice the percent alcohol content by volume measured at a temperature of 60°F. Grain alcohol, or pure ethyl alcohol, is 200 proof. A 90-proof whiskey is 45% alcohol. European countries use a simpler version of proof in which the percent alcohol by volume *is* the proof measure. This is called *Gay-Lussac proof,* and is listed in *degrees* instead of percents.

In Great Britain, proof is based on the concept of *proof spirit.* A 100-proof spirit is defined to be 48.24% alcohol by weight, or 57.06% alcohol by volume. Higher or lower alcohol content is then designated in "degrees above or below proof." To convert British proof to American, multiply by 8 and divide by 7. Thus a 70-proof liquor in the British system would be 80-proof American or 40-proof Gay-Lussac. In the British system, 70-proof might also be referred to as "30 under proof," meaning that it is 30 percentage points below the standard of 100 proof (which is about 57% alcohol).

Φ ———————————————————————————————— Φ

FIREWATER

The origins of the term "proof-spirit" are obscured by the many conflicting explanations offered by normally dependable reference books. It is apparently a historical oddity that historians have not thought important enough to pin down. But this much is certain: the idea for assessing the alcohol content of distilled liquors came from tax collectors. Because such beverages often were watered down, the volume of liquid was seen as less important than the volume of alcohol. But how to measure it? All of the books agree on this point as well: the first test for "proof" involved gunpowder.

Alcoholic beverages consist almost entirely of alcohol and water, with just a trace (although a crucial one) of flavoring substances. Alcohol can be ignited, while water cannot. Gunpowder can also be ignited. This gave someone the idea of testing how much water a liquor contained by mixing it with gunpowder to see how well it would burn. In some accounts, the gunpowder was dissolved in the liquor; in others, it was merely soaked with liquor. The most delicate part of the operation was the igniting of the mixture. If it burned steadily, or if the gunpowder would still ignite after the alcohol had burned off, this was considered "proof" that the liquor had not been diluted. (Some of the details are hard to come by, and whether this is the actual origin of "proof" is still questionable.) In what must have been an exciting series of trials and errors, it was determined that a mixture of about 57% alcohol by volume was the critical point below which the mixture would not burn, so this was assigned a value of 100-proof spirit. Mixtures above and below proof came to be specified in terms of "degrees" rather than "percent" (for example, "20 degrees below proof").

The rest of the story is perhaps best left to the imagination, because it is clearly something not to try at home.

Φ ———————————————————————————————— Φ

For beer and wine, alcohol content is usually measured in percent rather than proof. This measure, called *percent alcohol by volume*, is not the same as *percent alcohol by weight* for all liquids. Because alcohol is lighter than water, the percent alcohol by volume will always exceed the equivalent percent alcohol by weight.

By law, all European Union countries require the alcohol content of beers, wines, and spirits to be displayed in terms of percent alcohol by volume. In the United States, beer is sometimes labeled in terms of percent alcohol by weight, which is lower than percent by volume. For example, 3.5% by weight is equivalent to about 4.4% by volume. But because some states ban the display of alcohol content, most nationally distributed U.S. beers are not labeled at all.

Table 3.3 shows the percent of alcohol by volume for types of beer, wine, and spirits.

Table 3.3. Alcohol Content of Distilled and Fermented Drinks

Type	% alcohol by volume
Whiskey	40
Gin	40
Rum	40
Sherry	20
Port	20
Madeira	20
Table wines (white, red, rosé)	7 to 15
Sparkling wines	7 to 15
Lager beers	3 to 5
Top-fermented beers (ales, porters, stouts)	4 to 6.5
Malt liquor	8 to 15

2. Liquor Measures Beer, wine, and liquors come in a dazzling array of servings and container sizes. There are few if any universal standards. The base unit for a serving of liquor is the shot, but smaller servings—a dram, a nip, a finger—are possible. Below, roughly ordered by size, is a short glossary of the serving or package sizes of various alcoholic beverages.

> *dram*   a small quantity, a sip.
>
> *nip*   a sip (as in "a nip of spirits"); in the case of ale, up to a half-pint serving; also a small (about 50 ml) bottle of liquor.

finger   a serving of liquor filling a small whiskey glass to the depth of a finger's width—about ¾ inch or 19 mm. A two-finger serving is filled to the depth of two fingers—about 1½ inches.

shot   the amount that pours out of a bottle with a practiced twist of the wrist; anywhere from 1 to 1½ ounces. The standard capacity of a shot glass is 1¼ ounces, or about 37 ml.

double   a drink consisting of two shots.

triple   a drink consisting of three shots.

pony   a 1-ounce glass of liquor (much like a shot); a 7-ounce bottle of ale or beer (note that in both meanings *pony* is similar to *nip*).

jigger   a small whiskey glass or cocktail measure containing 1½ ounces or 45 ml; similar to a shot.

fifth   a fifth of a gallon, or ⅘ quart; approximately 750 ml.

split   a half-size bottle, 6 to 8 fluid ounces; a quarter bottle of champagne or dessert wine.

flagon   one quart.

yard, yard-of-ale   a trumpet-shaped glass 3 feet long containing just over a half-gallon.

demijohn   a wicker-covered glass bottle holding anywhere from 2 to 12 gallons.

firkin   a quarter of a barrel.

keg   a small cask or barrel amounting to about 15½ gallons.

UNIT PRICING

Most supermarkets display a *unit price* for each item they sell, listing prices of goods by weight, volume, count, length, or liquid measure. By a simple formula:

$$\text{unit price} = \frac{\text{total cost}}{\text{total amount}}$$

♦ A 24-ounce box of cereal costs $3.79. The unit price, in cents per ounce, would be

$$\text{unit price} = \frac{379}{24} = 15.8 \text{ cents per ounce}$$

To convert to cents per pound, multiply this by the conversion factor 16 (since there are 16 ounces in a pound). The resulting unit price is 252.8 cents per pound, or about $2.53 per pound.

Φ ———————————————————————————— Φ

SALMANAZARS AND REHOBOAMS

In the realm of weights and measures, exotic names are routinely legislated out of existence. The old British monetary system, with its farthings and halfpence, was metricized into oblivion several decades ago. Less noticed, and initially less lamented, was the old system of naming champagne bottle sizes. Today, the only descriptive name that can legally appear on a champagne bottle is the word *magnum,* which in Latin means "big thing." But a magnum is far from being the biggest champagne bottle. It is based on the *standard bottle,* which originally held 800 milliliters but now contains only 750. This is the standard table-wine bottle size, and the magnum is twice as big.

Far more impressive are the names that no longer appear on labels. They include the *jeroboam,* the equivalent of 4 standard bottles, and the most common size used in christening ships; the *rehoboam,* equivalent to 6 bottles; the *methuselah,* equivalent to 8 bottles; the *salmanazar,* equivalent to 10 (sometimes 12); the *balthazar,* equivalent to 16; and finally the *nebuchadnezzar,* the equivalent of 20 standard bottles, which amounts to almost 15 liters.

The names are biblical. Jeroboam and Rehoboam were the first kings of Israel and Judah, respectively, after the division of Solomon's kingdom into two parts. Methuselah, the patriarch listed in Genesis, lived for 969 years. Salmanazar, Balthazar, and Nebuchadnezzar were all kings of Babylon around the time of Daniel.

Although changes in units of measure are usually instituted for practical, economic, and scientific reasons, none of these justifies retiring these patriarchs, who had been such good friends and purveyors of good cheer to so many.

Φ ———————————————————————————— Φ

For purposes of comparison, the units of measure are generally kept uniform for competing items, but there are exceptions. One brand may show a per-pound unit price while a competing brand shows a per-ounce price. For tips on conversions, look closely at the product labels: a 24-ounce box of cereal will probably be marked 1.5 pounds as well, which should remind you that there are 16 ounces in a pound. If you multiply the per-ounce price by 16, the per-pound price will result. In general, such comparisons can be made by looking closely at the product label, which will give the size or amount in at least two units.

◆ Assume that two jars of honey, one containing 12 ounces and the other containing 18 ounces, cost $1.99 and $3.24, respectively. The unit prices are given as 17¢/oz and $5.76/qt. The unit prices can be compared using the fact that there are 32 ounces in a quart. A much simpler approach is to compare the jar sizes: 18 is ³⁄₂ of 12. The 12-ounce jar's price can be rounded to $2. Multiply $2 by ³⁄₂ and you will find that the honey in the smaller jar would cost $3 for 18 ounces and is therefore a better buy. The larger package does not always have the smaller unit price.

Energy Sources

ELECTRICITY

The basic units used to measure electrical pressure, current, and resistance are *volts, amperes* (or *amps*) and *ohms*. To visualize how electricity is conducted through a wire, imagine the flow of water through a pipe. The *voltage* is equivalent to the water pressure, the *current* (in amps) to the rate of flow, and the *resistance* (in ohms) can be compared to the size or resistance of a waterwheel turned by the force of the water.

The voltage in any household outlet is either 120 volts (for standard outlets) or 240 volts (for heavy-duty outlets required by clothes dryers and air conditioners). (Although this figure should remain constant, electric companies occasionally have to decrease the voltage during periods of peak demand. During critical power shortages, this results in a brownout.)

Voltage and current can be multiplied together to find electric power usage. This section explains each of these units and shows how they relate to a monthly electric bill.

1. A *kilowatt-hour* (kWh) is the basic unit of electric power consumption used by utility companies to calculate electric bills. It is the amount of power used by a device that draws 1 kilowatt (that is, 1000 watts) over 1 hour. Every electrical appliance or device carries a label indicating the amount of power it consumes. A 75-watt bulb, for example, requires 75 watts of power to light. Given the wattage and the duration of use, the cost of operation can be calculated by estimating kilowatt-hour usage and multiplying by the usage charge indicated on an electric bill. (See the example below.)

Not all devices are labeled by wattage, however. Some are labeled by amperage, resistance, and even horsepower. But all of these can be converted to wattage through simple formulas.

2. Wattage The number of watts used by a device can be calculated in a number of ways. Once the wattage has been calculated, it can easily be converted to kilowatts, and then to kilowatt-hours by the following formula:

$$(\text{watts} \div 1000) \times \text{hours} = \text{kilowatts} \times \text{hours} = \text{kilowatt-hours (kWh)}$$

- A 100-watt bulb uses $^{100}/_{1000} = 0.1$ kilowatts. If it burns for 20 hours, it uses $(0.1 \text{ kW}) \times (20 \text{ hours}) = 2$ kWh.

- A string of 20 15-watt Christmas bulbs requires $20 \times 15 = 300$ watts of power. If the lights are turned on for 5 hours each day, they consume $(300 \text{ watts} \div 1000) \times 5 \text{ hours} = 1.5$ kWh per day.

3. Amperage to Watts Electric current is measured in units of *amperes* (or simply *amps*). A typical house has 100-amp electrical service, although many are now outfitted at 200 amps. The circuit-breaker box distributes this current through various circuits that may be rated at 15, 20, or 30 amps. If the total amperage on a circuit exceeds the rating, the circuit breaker will open to prevent the circuit from overheating.

Current (in amps) can be converted to power usage (in watts) by using this formula:

$$\text{power} = \text{voltage} \times \text{current}$$

Or in other words:

$$\text{watts} = \text{volts} \times \text{amps}$$

The standard household outlet supplies 120 volts, so converting from amps to watts involves multiplication by 120. For a heavy-duty outlet, multiply by 240 volts.

- A 10-amp toaster drawing 120 volts uses 120 volts $\times$ 10 amps = 1200 watts of power, or 1.2 kilowatts. If the toaster is used 15 minutes per day, then it is used $30 \times 15 = 450$ minutes or $7\frac{1}{2}$ hours per month. The number of kilowatt-hours used would be

$$(1200 \text{ watts} \div 1000) \times 7.5 \text{ hours} = 9 \text{ kWh}$$

4. Resistance Some appliances are rated by their electrical resistance, which is measured in *ohms*. A simple formula known as *Ohm's Law* relates voltage, resistance, and current. It says:

$$\text{amps} = \text{volts} \div \text{ohms}$$

◆ The toaster described above has a resistance of 12 ohms. When plugged into a 120-volt circuit, according to Ohm's Law,

$$\text{current} = \frac{120 \text{ volts}}{12 \text{ ohms}} = 10 \text{ amps}$$

◆ A microwave oven with a resistance of 8 ohms operating on a standard 120-volt circuit uses 120 volts ÷ 8 ohms = 15 amps.

As described above, once the current (in amps) is known, the power in watts can be calculated using the formula *watts = volts × amps*. Combining this with Ohm's Law produces two formulas that can be used to convert (1) amps and ohms to watts, or (2) volts and ohms to watts.

$$\text{watts} = \text{volts} \times \text{volts} \div \text{ohms}$$

$$\text{watts} = \text{amps} \times \text{amps} \times \text{ohms}$$

◆ The power used by the microwave oven in the previous example can be calculated using either of the above formulas, where the voltage is 120, the amperage is 15, and the resistance is 8 ohms:

$$\text{amps} \times \text{amps} \times \text{ohms} = 15 \times 15 \times 8$$
$$= 1800 \text{ watts, or } 1.8 \text{ kW}$$

The voltage supplied to household outlets can vary, generally between 110 and 120 volts, but for purposes of estimating power usage it should be thought of as constant. This makes power usage directly proportional to current. By Ohm's Law, the current (amperage) is inversely proportional to the resistance (ohms). (For an explanation of proportionality, see Chapter 2.) Thus the power usage increases as the resistance falls; an appliance with low resistance draws more power than one with high resistance. Telephone wires, for example, have very high resistance and consequently draw little power. Some low-power systems use a low-voltage transformer or power adapter to convert 120 volts to about 24 volts or less. Such a transformer might be used to power a doorbell, an electric train set, or a laptop computer.

• A laptop computer has an adapter that converts 120 input volts to 24 output volts. It is rated at 1.04 amps. Its power usage is found by multiplying voltage times current:

$$24 \text{ volts} \times 1.04 \text{ amps} \approx 25 \text{ watts}$$

5. Horsepower to Watts Electric power is sometimes measured in *horsepower*, especially in connection with electric motors. One horsepower is the power required to lift a 550-pound weight 1 foot in 1 second. It is equivalent to 746 watts of power. To convert horsepower to watts, multiply by 746.

• Assume a table saw has a 2.5-horsepower motor. The wattage required to run the saw would be 2.5 × 746 = 1865 watts. If the saw runs for 15 minutes, its power usage would be

$$(1865 \text{ watts} \div 1000) \times 0.25 \text{ hours} \approx 0.47 \text{ kilowatt-hours.}$$

6. Electric Meters There are several types of electric meters. Some have digital readouts, but older models use four or five dials to indicate kilowatt-hour usage in units of 1, 10, 100, 1000, and 10,000 kWh. The dials alternate in their direction of numbering; one is numbered clockwise from 0 to 9, the next counterclockwise from 0 to 9, and so on. The indicator on each dial should be read like an hour hand; that is, it indicates the digit most recently passed. A hand between the 5 and the 6, even if it is closest to the 6, indicates 5.

Digital or dial meters show a cumulative total of kilowatt-hour usage. The dials are read from left to right to obtain a number that can be checked against the reading for the previous month to find the total usage for each month.

31,641 kilowatt-hours

7. Cost of Electrical Usage A monthly electric bill shows the number of kilowatt-hours used in a month, and perhaps the average household usage per day. Most of that usage usually comes from heating and air-conditioning systems and large appliances. By calculating power usage, it is possible to estimate the monthly cost of operating any household electrical device, something an electric bill does not itemize. This task is simplified in the case of large appliances, which are required to display an energy-efficiency rating and the estimated annual cost of operation. For smaller appliances, in those cases where the approximate usage can be estimated, operating cost can be determined directly. Table 3.4 gives the approximate levels of power consumption for a variety of large and small household appliances. These can be used to estimate each appliance's contribution to a monthly or annual electric bill.

Table 3.4. Power Consumption of Household Appliances (in watts)

Appliance	Approx. capacity
Water heater	2500
Air conditioner	1500
Washing machine	1000
Clothes dryer	4400
Iron	1100
Kitchen range	8000
Refrigerator	240
Dishwasher	1750
Toaster	1150
Hair dryer	1600
Television	200
Radio	50
Lamp	60–100
Overhead light	150
Sunlamp	280

◆ An electric clothes dryer runs on a 220-volt outlet. On a 20-amp circuit, the power usage would be

$$(220 \text{ volts}) \times (20 \text{ amps}) = 4400 \text{ watts}$$

A load requiring 45 minutes to dry would use

$$(4400 \text{ watts} \div 1000) \times (.75 \text{ hours}) = (4.4 \text{ kW}) \times (.75 \text{ hours})$$
$$= 3.3 \text{ kWh}$$

At about 9.5 cents per kilowatt-hour, this would cost

$$(3.3) \times (\$0.095) = \$0.3135, \text{ or about 31 cents}$$

At a rate of four loads each week, the annual cost of operation would be

$$(4 \text{ loads per week} \times 52 \text{ weeks}) \times (\$0.31 \text{ per load}) = \$64.48$$

The hourly cost of operating the dryer is found by multiplying its wattage (4.4 kW) times the hourly cost of power (in dollars per kilowatt-hour).

(4.4 kW) × ($0.095/kWh) × (1 hr.) = $0.418, or approx. $0.42/hr.

8. Sliding Scales Electric bills are generally calculated on a sliding scale. The first 100-200 kilowatt-hours are the most expensive (perhaps 9 or 10 cents per kWh), but after that the rate drops considerably.

- An electric bill indicates a monthly usage of 715 kWh. The first 100 units are priced at $0.098, the next 100 at $0.072, and anything over that at $0.062. The cost of electrical usage would be

$$100 \times (0.098) = \$9.80$$
$$100 \times (0.072) = \$7.20$$
$$\underline{515 \times (0.062) = \$31.93}$$

| Total | $48.93 |

(An additional monthly service charge would raise the bill's total somewhat.)

NATURAL GAS

1. Therms and BTUs Most gas companies measure gas usage in units of *therms*. A therm is about 1000 cubic feet of gas, although this can vary. Gas meters measure usage in cubic footage, but billing is done on the basis of therms. Consequently a gas bill will sometimes indicate a *conversion factor* from units of 1000 cubic feet of gas to therms of gas (e.g., "Therm factor: 1.031"). The two measures are always very close.

A therm of gas converts to 100,000 BTU (British Thermal Units). One BTU is the amount of heat required to raise the temperature of 1 pound of water 1 degree Fahrenheit. Roughly speaking, one BTU is the amount of energy released by striking a wooden match.

2. To calculate a gas bill, the utility company reads the gas meter on the same day each month and subtracts the previous month's reading in order to find the total therms used during the month. Like an electric bill, the gas bill includes a monthly fee and employs a sliding scale. It might look something like this:

Monthly charge	$5.70
First 30 therms	.6523
Additional therms	.3871

If the customer used 47 therms during the month, the bill would look like this:

Oct 17 meter reading	6659
Sept 17 meter reading	6612
Therms billed	47

Monthly charge	$ 5.70
30 therms @ .6523	$19.57
17 therms @ .3871	$ 6.58
Total	$31.85

3. Degree-Days A *degree-day*, a measurement used in estimating heating requirements for a home or office building, is a unit representing the difference between the average daily temperature and 65°F (a reference temperature). The average daily temperature is found by averaging the high and low temperatures recorded for a given day.

⬥ One day the temperature ranged from a low of 25°F to a high of 37°F. The number of degree-days is calculated as

$$65 - \frac{25 + 37}{2} = 65 - 31 = 34 \text{ degree-days}$$

The degree-days for each day in a heating season are usually summed to find a seasonal total. Days that average above 65°F during the heating season are ignored. Because the fuel usage of a house is proportional to the number of degree-days in a heating season, power companies and local weather bureaus maintain records of monthly and seasonal totals, and these can be used to calculate the heat loss in a house using the following formula:

BTU heat loss = heat rate of house × 24 × degree-days

100,000 BTU equals 1 therm; therm units are itemized on gas bills. The *heat rate* of a house is the hourly heat loss (see page 138) divided by the difference in indoor and outdoor temperature.

Construction

CALCULATING SQUARE FOOTAGE AND CUBIC FOOTAGE

In order to estimate the amount of wallpaper, flooring, tile, paint, carpet, roofing, or driveway sealer needed for a job, it is necessary to calculate the square footage to be covered.

Contractors almost never estimate materials by calculating in their heads. A few experienced contractors may "eyeball" the job and order material without making any written computations, but most make use of simple slide-rule calculators provided by building material suppliers, which give quick and accurate totals based on the dimensions of the job. Lumberyards use a modified yardstick to calculate board feet, and there are also calculators for asphalt, concrete, roofing and siding shingles, and insulation. With a concrete calculator, for example, the user lines up the measured length, width, and thickness dimensions, and then reads off the number of cubic yards required for the job. All of these carry out simple calculations based on the formulas outlined below.

In this section the terms *feet (foot), inch(es),* and *by* will be used interchangeably with the symbols ', ", and ×.

1. Household square footage Calculations for everything but circular areas are based on the areas of rectangles (area = length × width) and areas of triangles (area = base × height ÷ 2). Because of the complexity of most floor plans and wall profiles, it may be necessary to break irregularly shaped spaces (L-shaped rooms, for example) into two or more rectangles.

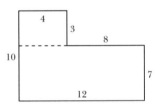

• The floor plan shown above can be broken into two rectangles: a 7' by 12' rectangle and a 3' by 4' rectangle. These have areas of 84 and 12, respectively, which add up to 96 square feet. Alternatively, we may think of the room as a 10' by 12' rectangle minus a 3' by 8' piece. By this method the area would be calculated as 120 (12' × 10') minus 24 (3' × 8'), or 96 square feet.

• If a room is 12 feet long and 14 feet wide and has 8-foot-high ceilings, its total wall space would be:

2 walls @ 8' by 12', or 2 × 96 = 192 sq. ft.
2 walls @ 8' by 14', or 2 × 112 = <u>224 sq. ft.</u>
Total   416 sq. ft.

A gallon of paint covers approximately 400 square feet (although this may vary considerably with different types of paint and stain—the exact figure is always given on the can). Assuming that there are door and window openings that can be subtracted from the total (as a rule of thumb, subtract 20 square feet for each door, and 15 square feet for each standard window), the surface area could safely be covered by a gallon labeled to cover 400 square feet.

2. Legal Square Footage The square footage of a house, as used in tax assessments or in descriptions used by realtors and contractors, is the total amount of interior living space. It does not include the basement, attic, or garage. If any of these are finished spaces, their square footage is usually noted separately from the figure for the house.

Total square footage is computed using the *exterior* dimensions of the house. This can be done from a blueprint or from measurements made and recorded onto a sketch of the footprint of the house (the outline of the ground area it covers). Some houses have simple rectangular footprints, but those that do not can usually be broken into rectangular sections by drawing dotted lines as shown below. Using the formula for the area of a rectangle, the total square footage for each level can be calculated easily.

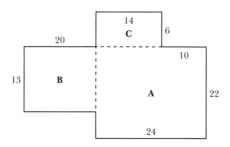

• The footprint of the house drawn here can be broken into three rectangles.

Area of A = 22 × 24 = 528 square feet
Area of B = 13 × 20 = 260
Area of C = 6 × 14 = 84
_____
1st-floor total   872 square feet

If the second level has the same footprint and there are no other levels, the total square footage would be double the first-floor total, which comes out to 1744 square feet.

3. Cubic Footage To calculate the *cubic footage* of a house, find the total area of the floor plan (or footprint) and multiply by the ceiling height.

If each floor has the same floor plan and the same ceiling height, multiply the cubic footage of the first floor by the number of floors. Otherwise, calculate the area of the floor plan for each level, and multiply by its ceiling height.

♦ A three-story house has identical first and second floors, each of which has 725 square feet of floor space and 9-foot ceilings. The third story has 450 square feet and 8-foot ceilings. The cubic footage can be calculated as follows:

1st floor: 725 square feet × 9 feet = 6,525 cubic feet
2nd floor: 725 square feet × 9 feet = 6,525 cubic feet
3rd floor: 450 square feet × 8 feet = 3,600 cubic feet
Total   16,650 cubic feet

UNITS OF MEASURE

1. Shingles Shingles are sold in units called *squares*. A square is equivalent to a 10′ × 10′ area, which totals 100 square feet. To estimate the number of squares needed for a job, find the square footage of the area to be covered and divide by 100.

♦ An 80-foot-long building has a simple pitched roof that measures 25 feet from gutter to peak. Thus the roof consists of two 25′ × 80′ rectangles. Each has a square footage of 25′ × 80′ = 2000, for a total of 4000 square feet. Dividing by 100 to find the number of squares, we get $^{4000}/_{100}$ = 40 squares.

Roofs generally have rectangular surfaces, although many have irregularities such as dormers. Dormers are not difficult to estimate because they can be broken into triangles. In fact, the two sides of a dormer can be combined to form a rectangle.

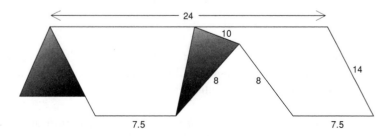

• The roof pictured above has a single dormer. The two roof surfaces of the dormer can be pieced together to make an 8′ × 10′ rectangle, which has 80 square feet of area. The dormer divides the roof into two principal surfaces that can also be arranged to form a rectangle—in this case with length 19.5′ and width 14′. The total area of this rectangle is 14 × 19.5, or 273 square feet.

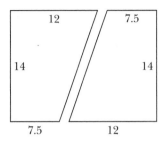 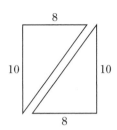

The total surface area of the dormer and the roof (on this side of the house) is 80 + 273, or 353 square feet. If the opposite side is identical, the job would require twice this amount, or 706 square feet. Seven squares of shingles would not quite be enough to do the job, especially if some allowance for waste is included (as it should be).

Extra shingles can be purchased in bundles, a *bundle* being usually a fourth or a third of a square, and thus covering anywhere from 25 to 35 square feet, depending on the type of shingle.

2. Roof Pitch The slope of a roof is given in terms of *pitch*, which describes the ratio of inches of rise to feet of run. Thus a pitch of 12, with 12 inches of rise to 1 foot of run, represents a 45° angle. Any pitch over 12 is steeper than 45°.

If the desired pitch of a roof is known, the length of the rafters can be calculated for any span. (Assuming a simple gable roof, as shown in the following figure, the span is the wall-to-wall width covered by the roof.) To find the rafter length: (1) square the pitch, (2) add 144, and (3) take the square root. (This gives the rafter length in inches per foot of run.) Multiply the result by the run (half of the span, in decimal feet) to get the rafter length in inches.

• Assume a 10-pitch gable roof is to span 24 feet. Each half of the roof will span 12 feet. To find the rafter length, square 10 to get 100, add 144 to get 244, and take the square root to get 15.6205. Now multiply by 12 feet to get 187.446 inches, or about 15'7½".

Rafters are usually cut long in order to provide overhang. Since the calculation above uses the interior span, the rafter length that results is an interior dimension.

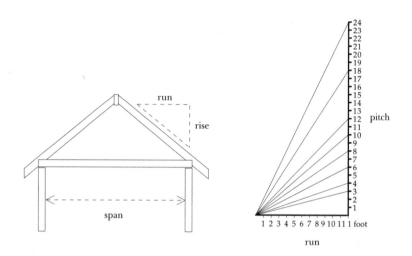

run

3. Board Feet Lumber, in the form of siding, flooring, studs, joists, and other boards, is sold by the board foot. A *board foot* is the equivalent of a piece of wood 12 inches long by 12 inches wide by 1 inch thick. The volume of a board foot is 144 cubic inches.

To calculate the number of board feet in a piece of wood, compute its volume (this may require converting its length from feet to inches) and divide by 144.

board feet = (length in inches) × (thickness in inches)
× (width in inches) ÷ 144

◆ A 10-foot 1″ × 4″ oak board has a length of 10 × 12 = 120 inches and a volume of (120″) × (1″) × (4″) or 480 cubic inches.

Dividing by 144 gives 3⅓ board feet. If the oak costs $2.75 per board foot, the cost of the piece would be

$$\left(3\frac{1}{3}\right) \times (\$2.75) = \$9.17$$

A simpler formula is:

board feet = (length in feet) × (width in feet) × (thickness in inches)

♦ A 12-foot 2″ × 10″ joist contains

$$(12) \times \left(\frac{10}{12}\right) \times (2) = 20 \text{ board feet}$$

(Notice that the 10″ width dimension is expressed as $^{10}/_{12}$ of a foot.)

1-inch, ⅞-inch, and ¾-inch stock are all calculated as 1 inch thick in the board-feet formula. To calculate board feet for thinner stock (½″ thick or less), use a thickness of ½″.

4. Lumber Dimensions Lumber comes in a few general classifications:
A *board* is wood that is milled less than 2 inches thick.
Dimension lumber can be 2 to 5 inches thick. This includes framing lumber such as 2 by 4's, joists, and planks.
Timbers are more than 5 inches in thickness and width. This includes most beams.
Millwork includes moldings, casing, tongue-and-groove, and flooring, and *sheathing* includes plywood and particleboard.
Prices of millwork and sheathing are set by the linear foot (or running foot), by the 4′ × 8′ sheet, or by the square foot. Other lumber prices are set by the board foot, which is calculated from the dimensions of the piece.
Although a board, a timber, or a piece of dimension lumber is finished, or *dressed,* to dimensions that are smaller than the named dimensions, it is the named dimension that is used in the calculation of board feet. A two-by-four, for example, is dressed to 1½ by 3½ inches. Yet the board-footage of a 12-foot two-by-four is calculated using a full 2 inches for the thickness and 4 inches for the width.

$$\text{board feet} = (12 \text{ feet}) \times \left(\frac{4}{12} \text{ foot}\right) \times (2 \text{ inches}) = 8 \text{ board feet}$$

Other dressed dimensions are given in Table 3.5.

Table 3.5. Lumber Sizes

Nominal size (in inches)	Actual size (dressed)	Board feet per linear foot (To convert to board feet, multiply length by:)
1 by 2	$\frac{3}{4} \times 1\frac{1}{2}$	$.1667 \left(\frac{1}{6}\right)$
1 by 4	$\frac{3}{4} \times 3\frac{1}{2}$	$.3333 \left(\frac{1}{3}\right)$
1 by 6	$\frac{3}{4} \times 5\frac{1}{2}$	$.5 \quad \left(\frac{1}{2}\right)$
1 by 8	$\frac{3}{4} \times 7\frac{1}{2}$	$.6667 \left(\frac{2}{3}\right)$
1 by 10	$\frac{3}{4} \times 9\frac{1}{2}$	$.8333 \left(\frac{5}{6}\right)$
1 by 12	$\frac{3}{4} \times 11\frac{1}{2}$	1
2 by 2	$1\frac{1}{2} \times 1\frac{1}{2}$	$.3333 \left(\frac{1}{3}\right)$
2 by 3	$1\frac{1}{2} \times 2\frac{1}{2}$	$.5 \quad \left(\frac{1}{2}\right)$
2 by 4	$1\frac{1}{2} \times 3\frac{1}{2}$	$.6667 \left(\frac{2}{3}\right)$
2 by 6	$1\frac{1}{2} \times 5\frac{1}{2}$	1
2 by 8	$1\frac{1}{2} \times 7\frac{1}{2}$	$1.333 \left(\frac{4}{3}\right)$
2 by 10	$1\frac{1}{2} \times 9\frac{1}{2}$	$1.667 \left(\frac{5}{3}\right)$
2 by 12	$1\frac{1}{2} \times 11\frac{1}{2}$	2

Some types of lumber are rough-cut, and their named dimensions are the same as their true dimensions. A 4″ × 4″ cedar fence post, for example, is a true 4 inches by 4 inches, whereas a finished, pressure-treated 4″ × 4″ post is dressed to 3⅝ inches by 3⅝ inches.

5. Cordwood Wood for home heating is measured in cords. A *cord* is the equivalent of a stack of wood 4 feet wide by 4 feet high by 8 feet long. Its total volume is 4 × 4 × 8, or 128 cubic feet. A *cord foot* is a stack 4 feet wide by 4 feet high by 1 foot thick. Thus there are 8 cord feet in a cord.

A fireplace is not a very efficient means of heating a house—too much of the heat goes out the chimney. But most modern wood-burning stoves are far more efficient, being designed to transfer as much of the heat of combustion as possible. With cordwood, there can be a considerable disparity between the heat generated by equal volumes. Hardwoods like birch and maple, for example, are denser than softwoods like fir, pine, and spruce, and can generate up to twice as much heat. The hardest woods generate about 30 million BTU per cord, the softest woods about 15 million BTU.

6. Concrete Concrete is measured and sold by the cubic yard (called simply a *yard*). A yard is a volume that is 3 feet long, 3 feet wide, and 3 feet high. Thus it contains $3 \times 3 \times 3 = 27$ cubic feet.

To estimate the amount of concrete needed for a job, it is necessary to find the cubic footage of the area to be filled.

✦ A 50-foot driveway, 10 feet wide, is to be poured to a depth of 6 inches. The volume is (50 feet) $\times$ (10 feet) $\times$ (½ foot) = 250 cubic feet. Divide this by 27 to obtain the number of cubic yards.

$$\frac{250}{27} \approx 9.3 \text{ yards of concrete}$$

If concrete is selling for $62 per yard, the job would cost $9.3 \times 62 = \$576.60$. To add a 5% sales tax, multiply the total by 1.05, for a total billed amount of $\$576.60 \times 1.05 = \$605.43$.

Concrete jobs are usually estimated using a *concrete calculator*—a kind of slide rule that transforms the measured dimensions (length, width, thickness) into the number of yards required. Any concrete supply company can provide one.

7. Wire Sizes Different thicknesses of wire are used for circuits carrying different amounts of current. The wire thicknesses are given as gauge numbers, which run from 0 to 14, where 0 is the largest diameter and 14 is the smallest. No. 2 or 3 wire is used to provide service from the street to a house equipped with 100-amp service. No. 10, 12, and 14 wire are used in wiring the house. A No. 14 wire is recommended for 15-amp circuits meant for lighting only. No. 12 wire with a 20-amp breaker is used on circuits meant for small appliances and general utility outlets; No. 12 wire may also be used on 15-amp circuits. Thick No. 10 wire is used for heavy-duty, 30-amp circuits, and can be used with 110–120 or 220–240 volts.

8. Nail Sizes The use of the word *penny* to describe sizes of nails has a disputed origin. According to one theory, *penny* means *pound*, and the penny measure is the weight in pounds of 1000 nails. Thus fivepenny nails would weigh 5 pounds per thousand. Another explanation—more convincing because it seems a more natural way to sell nails—holds that the word *penny* originally referred to the cost in pennies per hundred nails. Thus a fivepenny nail would have cost 5 cents per hundred, an eightpenny nail 8 cents per hundred. Whatever its origin, *penny* is now a measure of length.

Table 3.6 shows standard wire nail lengths with counts per pound. The abbreviation for *penny* is the letter *d*. With nail sizes, there is far more agreement as to length than to count per pound, but even the length measurements given here should not be relied on as universal standards.

Table 3.6. Nail Sizes

		Approximate count per pound		
Designation	Length (inches)	Common	Finish	Box
2d	1	900	1350	1000
3d	1¼	615	850	650
4d	1½	325	600	450
5d	1¾	265	500	400
6d	2	200	310	225
7d	2¼	160	—	—
8d	2½	105	190	140
9d	2¾	90	—	—
10d	3	75	120	90
12d	3¼	60	110	85
16d (spike)	3½	45	90	70
20d	4	30	60	50
30d	4½	20		
40d	5	14		
50d	5½	10		
60d	6	8		

A *keg* of nails is a 100-pound container.

9. Insulation Requirements Every building material has an *R-value*, which is a number indicating its ability to retain heat (R stands for *resistance*). The highest R-values belong to insulating materials such as fiberglass batting and polystyrene. R-values are given in units per inch or per meter. Table 3.7 shows R-values for some common building and insulating materials.

Table 3.7. R-Values for Common Building Materials

Insulator	R-value
Film of air on outside wall	0.2
Film of air on inside wall	0.7
Airspace in wall	1.0/inch
Wood siding	0.8/inch
Plywood	1.25/inch
Wallboard	1.0/inch
Asphalt shingles	0.45/inch
Brick	0.2/inch
Fiberglass batting	3.17/inch
Fiberglass loose fill	3.0/inch
Perlite fill	2.75/inch
Extruded polystyrene	5.0/inch
Glass pane	0.9
Insulated glass:	
$\frac{3}{16}''$ air space	1.4
$\frac{1}{4}''$ air space	1.5
$\frac{1}{2}''$ air space	1.7
$\frac{3}{4}''$ air space	1.9

To find a wall's total resistance to heat flow, multiply the thickness of each component by its R-value and sum them up.

Film of air on outside wall	0.2
Film of air on inside wall	0.7
Wood siding (0.8/in × 0.5 in)	0.4
Plywood sheathing (1.25/in × 0.5 in)	0.6
Fiberglass batting (3.17/in × 3.5 in)	11.1
Wallboard (1.0/in × 0.5 in)	0.5
Total	13.5

To calculate the BTU heat loss per hour, find the surface area of the wall (in square feet), multiply by the difference between indoor and outdoor temperature (in degrees Fahrenheit), then divide by the R-value. The total for the walls can be added to the total for the roof, for each window, and for each door. (See also discussion of degree-days on page 000.)

♦ A 3' × 5' double-pane window with $\frac{1}{2}''$ air space has a surface area of 15 square feet and an R-value of 1.7. The outdoor temperature is 25°F and the indoor temperature is 65°F. The window's BTU loss per hour is

$$15 \times (65 - 25) \div 1.7 \approx 350 \text{ BTU}$$

Automobiles

When shopping for, purchasing, driving, and maintaining a car, the consumer is confronted with a confusion of numbers that can easily be taken for granted or simply ignored. Yet numbers are the basis of informed choices. Some of them, such as engine specifications, are of interest mostly to car buffs. But others, such as acceleration, stopping distance, horsepower, gas mileage, and tire numbers, are crucial indicators of performance and can show how well a car will serve the buyer's needs. The purpose of this section is to help make sense of the numbers and to explore a few basic automotive principles that the seller rarely bothers to explain.

1. The *speed* or *velocity* of a car is a measure of its change in position with respect to time. For any trip a car makes, its *average velocity* is calculated as the ratio of total distance traveled to the time the trip required.

$$\text{average velocity} = \frac{\text{distance traveled}}{\text{elapsed time}}$$

(Notice that this is equivalent to the familiar algebraic formula: rate times time equals distance.)

- A car makes a trip in 6.4 hours. The odometer reading is 65,488 at the start of the trip and 65,744 at the end. The average velocity of the car on the trip was

$$\frac{\text{distance traveled}}{\text{elapsed time}} = \frac{65744 - 65488}{6.4} = \frac{256}{6.4} = 40 \text{ miles per hour}$$

Instantaneous velocity is simply the speed of the car as indicated by the speedometer at any instant of time. The instantaneous velocity can be thought of as the average velocity over a very short time interval.

- If a car travels a distance of 88 feet in 1 second, its average velocity (88 feet per second) is very close to its instantaneous velocity because the time interval is so short.

2. *Acceleration* is the rate of change of velocity. A car whose velocity is increasing (speeding up) is said to be *accelerating*. A car whose velocity is decreasing (slowing down) is *decelerating*. A falling object, for example, has an acceleration of 32 feet per second per second. This means that its velocity increases by 32 feet per second with each elapsed second. At 1 second, its velocity would be 32 feet per second; at 2 seconds, 64 feet per second; at 3 seconds, 96 feet per second; and so on.

The acceleration of an automobile can be estimated by taking speed readings at regular time intervals. The standard measure of acceleration is the time it takes a car to go from 0 to 60 miles per hour. If a car accelerates at 6 miles per hour *per second,* it will go from 0 to 6 mph in 1 second, 0 to 12 mph in 2 seconds, and ultimately, 0 to 60 mph in 10 seconds. This turns out to be rather slow, and most cars do somewhat better. Table 3.8 shows acceleration times for a Chevrolet Corvette and a Honda Accord.

Table 3.8. Acceleration: Chevrolet Corvette vs. Honda Accord Coupe (1998)

Acceleration	Corvette	Accord
0 to 30 mph	2.1 seconds	2.9 seconds
40	2.8	4.6
50	3.8	6.4
60	4.9	8.5
70	6.2	11.7
80	7.9	15.0
90	9.8	19.1
100	11.9	25.1
110	14.6	33.0
120	17.4	46.7
130	20.9	
140	28.7	
150	37.6	

Source: *Car and Driver,* October 1997

3. The gears of a car's *transmission* (or *transaxle* in a front-wheel-drive car) transform the rotational speed of the crankshaft into an acceptable range of rotation of the wheels. First gear is used to accelerate from a stop to about 10 miles per hour, at which speed the engine has reached several thousand revolutions per minute (rpm's). Second gear is used for the next range of speeds, and so on up to as many as five forward gears.

This transfer of rotational speed and rotational force is regulated by gear ratios. A *gear ratio* is the ratio of the number of teeth on a drive gear to the number of teeth on the gear it drives. This number determines how the rotational speed and rotational force (or *torque*) will be transferred. A drive gear with half as many teeth as the driven gear needs to rotate twice in order to cause one rotation in the driven gear. The torque (or twisting force) of the driven gear will be twice that of the drive gear. When a gear ratio reduces the rotational speed from one gear to the next, it also increases the rotational force by an equal factor. (See page 94 for more on gear ratios.)

In their annual rating of new automobiles, *Consumer Reports* uses two key indicators of engine performance, *final-drive ratio* and *engine revolutions per mile*. Both numbers are based on the performance of the car in high gear. The *final-drive ratio* is the number of engine revolutions for each revolution of the wheels. *Engine revolutions per mile* is the same as revolutions per minute at 60 miles per hour. (This is because 60 miles per hour is equivalent to 1 mile per minute.)

A low value for either measure relative to other models in the same class generally means better gas mileage and lower engine noise but less accelerating power. Table 3.9 compares two cars of very different classes, the Corvette and the Accord; relative to each other, their final-drive ratios reveal little.

4. Engine size is measured by the *displacement* of its cylinders. This is directly related to the amount of power it generates. The power derives from the combustion of fuel in the cylinder, which causes a piston to move down (and then back up). The pistons are attached to a crankshaft in a way that transforms their up-and-down motion into the rotational motion of the shaft. This is ultimately transferred to the wheels through the transmission and the drive train.

The displacement of a cylinder is the volume through which the piston moves between its highest and lowest positions in the cylinder. This volume is determined by the length of travel of the piston (the *stroke*) and the diameter of the cylinder (the *bore*). Since an engine cylinder is shaped like a simple tin can—that is, the shape technically known as a *cylinder*—its volume can be found by multiplying its height (or stroke) by the area of its circular cross-section. The formula (which is the standard formula for cylindrical volumes, given in the Appendix) is

$$volume = \pi \times r^2 \times h$$

where the radius r is half of the bore, and the height h is the stroke.

• A 1998 Chevrolet Corvette convertible has a stroke of 3.62 inches and a bore of 3.90 inches. The cylinder radius is half of the bore—about 1.95 inches. Thus

$$
\begin{aligned}
\text{displacement of one cylinder} &= \pi \times r^2 \times h \\
&= \pi \left(\frac{\text{bore}}{2} \right)^2 \times (\text{stroke}) \\
&= (3.1416) \times (1.95 \text{ in})^2 \times (3.62 \text{ in}) \\
&= 43.24 \text{ cubic inches}
\end{aligned}
$$

The *total displacement* is the displacement of all the engine's cylinders. This is expressed in cubic inches, cubic centimeters, or liters. The Corvette's total engine displacement is found by multiplying one cylinder's displacement by 8.

total displacement = (8 cylinders) × (43.24 cubic inches) ≈
346 cubic inches

The conversion factor from cubic inches to cubic centimeters (given in the conversion table on page 144) is 16.387. Thus

346 cubic inches = (16.387 × 346) cc ≈ 5670 cc, or about 5.7 liters

The displacement of an American automobile may be given in cubic inches or liters. A high-end 8-cylinder car will often exceed 300 cubic inches. A compact, such as the Honda Accord described in Table 3.9 below, will usually not exceed 140 cubic inches.

One way manufacturers use engine-displacement numbers is to differentiate among types of motorcycles. The number of cubic centimeters is often made part of the name—"Honda 450" or "Triumph 650," for example. A 250-cc engine belongs to a small bike, possibly a dirt bike, while 1000 or more cc's is the mark of a large road bike.

Table 3.9. A Comparison of Engine Specifications

	Corvette	*Accord*
Fuel economy	20 mpg avg.	25 mpg avg.
Cylinders	8	4
Cylinder bore	3.90 in (99.0 mm)	3.39 in (86.0 mm)
Stroke	3.62 in (92.0 mm)	3.82 in (97.0 mm)
Displacement	346 in^3 (5665 cc)	138 in^3 (2254 cc)
Final-drive ratio	3.42 to 1	4.06 to 1
Compression ratio	10.1 to 1	9.3 to 1

Because car manufacturers in the United States have not entirely adopted the metric system, the comparison of engine specifications of various makes and models requires some conversion formulas:

Liters to cubic inches:	Multiply by 61
Cubic centimeters to cubic inches:	Divide by 1000, then multiply by 61
Cubic inches to cubic centimeters:	Divide by 61, then multiply by 1000
Cubic inches to liters:	Divide by 61
Liters to cubic centimeters:	Multiply by 1000
Cubic centimeters to liters:	Divide by 1000

5. Compression Ratio The *compression ratio* of a cylinder measures the amount the fuel mixture is compressed before combustion. The fuel starts to enter the cylinder when the piston is at the bottom of its stroke. The fuel is compressed as the piston moves toward the top of the stroke. The space into which the fuel is compressed when the piston is at the top of the cylinder is referred to as the *combustion chamber.* If D is the cylinder displacement and V is the volume of the combustion chamber, then

$$\text{compression ratio} = \frac{D + V}{V} = \frac{\text{maximum chamber volume}}{\text{minimum chamber volume}}$$

A typical compression ratio for passenger cars is about 8 to 1 or 9½ to 1. The ratios for special-performance cars fall into the range of 11 to 1 up to 13 to 1.

6. Horsepower Engine output is most often measured in units of horsepower. One *horsepower* is defined to be the power required to lift a 550-pound weight 1 foot in 1 second. This standard was established by James Watt, a pioneer in steam-engine design, for whom the electrical unit of power is named. To rate his steam engines, Watt decided on a measure equivalent to 33,000 foot-pounds per minute, which is the power capacity of about 1½ real horses. An engine's horsepower is related to its displacement, but the way engine power is tested leads to several categories of horsepower.

The most meaningful category is *brake horsepower (bhp)*, which gives the best measure of the amount of power the engine supplies to the driveshaft as well as to all of the accessories, such as the air conditioner and power-steering mechanism. Brake horsepower is found by placing a braking device on an engine's output shaft and measuring the resistance at high rpm. While there is no exact conversion formula for auto-

Φ ———————————————————————————— Φ

HOW POWERFUL IS A HORSE?

James Watt, Scottish engineer extraordinaire, did not invent the idea that the standard measure of power should be based on the horse. But he did define the measure that we use to this day, and will probably continue to use even though it is not metric. The name Watt chose, *horsepower*, is simply too appealing to relinquish.

Watt's method of deriving one horsepower involves the formula for the circumference of a circle and the relationship between force, distance, and work. In Watt's day, horses and mules provided much of the mechanical power needed in mills by pushing or pulling a lever arm attached to a capstan—a hub that turned a shaft that drove the millstone. Imagine the circular path of the horse and a standard 12-foot lever, which forms the radius of the circle. The circumference of the circle, which is the distance traveled by the horse in one lap, is given by the formula $2\pi r$ (see Appendix), which works out to about 75.4 feet. Watt calculated that an average horse completes 144 laps in one hour, or 2.4 laps per minute. Since each lap is 75.4 feet, 2.4 laps per minute comes out to 75.4 × 2.4, or about 181 feet per minute. The resistance in the capstan, as estimated by Watt, required the horse to pull with a constant force of 180 pounds. Using the physics formula which says that "work equals force times distance," Watt calculated the work done as: (181 feet per minute) × (180 pounds) = 32,580 foot-pounds per minute, which he rounded off to 33,000 foot-pounds per minute. This, he decided, should constitute one horsepower.

One horsepower can be pictured as the power required to raise a 33,000-pound weight one foot in one minute, or to raise a 550-pound weight one foot in one second. The equivalent amount of electric power, in metric units named for the inventor, is 746 watts.

Φ ———————————————————————————— Φ

mobile engines, 1 liter of engine capacity normally converts to between 50 and 70 bhp. (1 liter is 1000 cc.) Higher ratios are attained by turbocharged engines, which can generate as much as 150 bhp per liter of displacement. Motorcycle engines can also achieve high performance; a

Honda 1100-cc motorcycle, for example, has an output of 175 bhp for its 1.1 liters.

Another measure of engine power, though a less useful one, is *indicated horsepower*, which is based on the pressure developed in the cylinders. Indicated horsepower exceeds brake horsepower by about 10%, because it does not correct for the reduction in output caused by the internal friction in the engine.

Another dubious measure is the advertised horsepower, or *manufacturer's horsepower*. This, like all power ratings, varies with the speed at which the car is tested. Since an engine's maximum horsepower is generated near its maximum rpm, manufacturers list the highest figure they can generate, labeling it the *advertised net horsepower at rpm*. A racing car, for example, generates its highest horsepower in the highest 1500 rpm of its range (which may go as high as 7000). A sedan (such as a 4-cylinder Toyota Camry, with its 2-liter engine) may be listed at about 120 horsepower, which it generates at top speed.

7. Car Sizes The automotive industry has developed four categories to describe the sizes of the cars they manufacture. They are based on the car's weight, its wheelbase (the distance between the wheels, center to center), and its overall length; see Table 3.10. Unofficially, the dealer markup on cars is 10–15% for subcompact and compact, 18–20% on midsize, and 22% on full-size. On luxury cars, vans, and pickups, the markup is usually 25%.

Table 3.10. Car Sizes

Size	Weight in lbs.	Wheelbase	Overall length
Subcompact	under 2500	under 100″	under 175″
Compact	2500–3000	100–105″	175–185″
Midsize	3000–3500	105–109″	185–200″
Full-size	over 3500	over 110″	over 195″

8. Gas Mileage A car's gas mileage is the average ratio of miles traveled to fuel consumed. Because driving conditions affect gas mileage, it is best to make this calculation using a full tank of gas, and to repeat the calculation many times in different weather and traffic conditions.

To find the gas mileage, fill the gas tank and zero out the trip odometer. After running the car somewhere near empty, refill the tank and record the number of gallons it takes to refill. Divide the odometer mileage by the number of gallons, and record the result. Repeat this for several tanks of gas.

◆ A car travels 364 miles on 13.45 gallons of gas. The gas mileage for that tank is 364 ÷ 13.45, or about 27 miles per gallon.

9. Oil Viscosity *Viscosity* is a measure of a liquid's ability to resist flow. High-viscosity oils are thick and flow slowly; low-viscosity oils are thin. Motor oils come in several grades for use with different cars under a variety of conditions. In the United States, the Society of Automotive Engineers (SAE), along with the American Petroleum Institute (API), has developed a rating system referred to as SAE numbers. These indicate the rate at which different oils will flow at different temperatures. The testing temperatures used—a range of values intended to simulate winter starting conditions as well as running temperatures—allow the oil to be rated at realistic extremes.

There are eleven SAE viscosity grades in all, six of which are designated "W," or winter, ratings. The thinnest winter grade is 0W, with thicker grades numbering 5W, 10W, 15W, 20W, and 25W. The high-temperature grades run from 20 to 60 in increments of 10. In both cases a higher number indicates a thicker oil.

Multigrade engine oils combine the effectiveness of low-temperature and high-temperature oils. Thus the low- and high-temperature ratings are combined into a dual rating such as 5W30. The numbers themselves are codes, not indexes, and thus have no precise relationship to one another. Because different viscosity tests are used for low and high temperatures, the two ratings do not refer to the same scale of viscosity. 10W40, for example, has an SAE rating of 10 at winter temperatures (for easier starting), and performs the same as an SAE 40 grade oil at running temperatures. The 10W40 designation indicates an oil that will resist thinning at high temperatures and thickening at low temperatures.

Some oils are rated only at low or high temperature, depending on their intended use. A 30-grade engine oil is not rated for low-temperature performance, but it will perform the same as a 5W30 oil at running temperature. A 5W-grade engine oil is not rated for high temperatures, but it will perform the same as a 5W30 at low temperatures.

10W40 is considered a year-round grade for moderate climates. A winter oil appropriate for an especially cold region might be a 5W30, which would allow for easier starting on cold mornings. In choosing which oil is appropriate, the consumer should always refer to the recommendation in the owner's manual.

10. Tire Sizes The numbers printed on the sides of automobile tires are mandated by the U. S. Department of Transportation. They refer not only to tire size but to certain performance tests. None of these indicators are self-explanatory. The standardized designations for tire sizes, for example, display a curious mixture of metric and U. S. systems of measurements. Three numbers are used to indicate the *tire width* (in millimeters), the tire's *cross-sectional aspect ratio* (a percent), and the *rim diameter* (in inches).

The *tire width* is the distance from sidewall to sidewall when seen from the front. This assumes a properly inflated tire that is not bearing the weight of the car.

The *aspect ratio* is a percent ratio of (1) the distance from the bottom of the rim to the road (also known as the *sidewall height*) to (2) the tire width.

The *rim diameter* is the inner diameter of the tire, which is the same as the outer diameter of the rim.

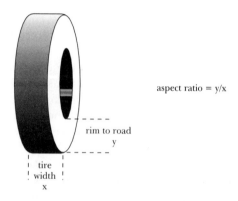

aspect ratio = y/x

rim to road
y

tire
width
x

Tire sizes also include codes designating the type of tire and its speed rating.

- In a tire size given as **P205 / 70 R 14**, P stands for *passenger* car and 205 is the tire width in millimeters. 70 is the aspect ratio, indicating that the sidewall height y is 70% of the tire width x. R stands for *radial* tire, and the last number indicates that the tire fits a rim with a diameter of 14 inches.

Other numbers that appear on the tire are:

The *load index* and *speed rating*. In a load/speed rating such as 87S, the 87 is a coded number that indicates a maximum weight the tire can carry at a speed designated S. The coded speed rating may appear with the load index number, or it may appear in front of the R in the tire size. A rating of S is recommended for passenger cars. A tire salesman may try to push a sports-car owner toward a higher rating (which will increase the price), but a rating in the vicinity of 85S is adequate for most cars and drivers. The most common speed ratings (most of them unrealistically high, thereby providing a margin of safety) are:

S: 106–112 mph
T: 112–118 mph
H: 118–130 mph
V: 130–149 mph
Z: 149 mph or more

Date of manufacture. The date of manufacture of any tire is coded in the last three digits of the DOT code printed on the side of the tire. Such a code might read: DOT M6 RV T1HR 428. The last three digits (428) indicate the 42nd week of 1998. (*Consumer Reports* cautions against buying a tire that is more than a few years old.)

The *tread-wear index.* This is intended to indicate how long the tread will last. It is an index number which rates every tire against a reference tire with a rating of 100. A rating of 400 means that a tire will last four times as long as the reference tire. An index in the 100s is considered low; near 500 is considered high. *Consumer Reports* notes that manufacturers conduct this test themselves, and there is no independent verification.

Traction and temperature rating. The traction rating is a test of straight-line stopping distance on wet pavement. The possible ratings are A, B, and C. According to *Consumer Reports,* the test is not very demanding, and about half of all passenger-car tires achieve a rating of A. The temperature rating gauges a tire's ability to withstand the heat generated under driving conditions. It is also an A, B, C scale, on which most tires rate at least B.

11. Braking Distance The time required for a car to come to a stop depends principally on four factors: (1) the reaction time of the driver, which averages ¾ of a second, (2) the speed of the car, (3) the condition of the tires and brakes, and (4) the condition of the road. The last two factors are hard to predict; but if both are assumed to be good, and the driver does not lock the brakes, then the first two factors will determine the stopping distance. For example, at 60 miles per hour (which is equivalent to 88 feet per second), a car will travel 66 feet in the ¾ seconds the driver takes to react. Assuming the car's brakes and tires work properly and the street is not slippery, the car will travel another 206 feet while the brakes are applied. Table 3.11 gives average values for braking distances at different speeds. Notice that the total stopping distance equals the reaction time plus the braking distance.

Table 3.11. Braking Distances

Speed (mph)	Reaction time (distance in feet)	Braking distance	Total stopping distance
10	11 feet	9 feet	20 feet
20	22	23	45
30	33	45	78
40	44	81	125
50	55	133	188
60	66	206	272
70	77	304	381

Source: *Automotive Encyclopedia*, rev. ed. (South Holland, Ill.: Goodheart-Wilcox, 1989).

As a rule of thumb, the proper distance for following behind another car at any speed should be at least two seconds. That is, a driver should be able to count off "one thousand one, one thousand two" before passing some landmark that the leading car has just passed.

4 The Physical World

Temperature

There are two types of temperature scales currently in use. One provides us with a convenient way to express indoor and outdoor temperatures, to measure body temperature, and to set our ovens. The other type of scale, called thermodynamic, is based on a physical limit called *absolute zero;* it was invented for scientific and engineering purposes and is rarely encountered outside of the laboratory.

There are metric and non-metric scales of each type. Fahrenheit and Celsius are our everyday scales, and the Rankine and Kelvin scales are their thermodynamic counterparts. Historically, a few other scales have come and gone, notably the Réaumur scale, which persisted in Europe into this century, and is now primarily associated with the brewing of beer.

1. The Celsius and Fahrenheit scales were originally based on the boiling and freezing points of water. In Celsius (basically identical to *centigrade*) this interval is divided into 100 units; in Fahrenheit, into 180 units. In both, the units are called *degrees,* but the degrees are not equal. A Fahrenheit degree is ⁵⁄₉ of a Celsius degree, a Celsius degree is ⁹⁄₅ of a Fahrenheit degree.

2. Temperatures convert from Fahrenheit to Celsius or vice versa by two simple conversion formulas. (Note that only the first uses parentheses, to indicate that 32 is first subtracted from the Fahrenheit temperature and that the *result* must be multiplied by ⁵⁄₉.)

$$C = \frac{5}{9} \times (F - 32)$$

$$F = \frac{9}{5} \times C + 32$$

- A temperature of 68°F converts to 20°C as follows:

$$C = \frac{5}{9} \times (68 - 32) = \frac{5}{9} \times 36 = 20$$

- A temperature of 15 °C is equivalent to 59°F:

$$F = \frac{9}{5} \times 15 + 32 = 59$$

- The boiling point of water is 212°F, which is equivalent to 100°C. The freezing point of water is 32°F, or 0°C.

3. The *Kelvin scale* uses Celsius degrees, setting 0 as *absolute zero*, the temperature at which the pressure of a gas is zero. The degree symbol is not used with the Kelvin scale; the unit of measure is the *kelvin*, abbreviated K. The relationship between Celsius and Kelvin is:

$$C = K - 273.15$$

$$K = C + 273.15$$

Thus the freezing point of water is 273.15 K, and the boiling point is 373.15 K.

4. The *Rankine scale* uses the same increments as the Fahrenheit scale. It is a thermodynamic scale in which absolute zero (−459.67°F) is set at 0° Rankine (abbreviated R). Thus

$$F = R - 459.67$$

$$R = F + 459.67$$

5. A fifth scale—the *Réaumur scale*—preceded the metric system (and thus the centigrade scale) in France and parts of Europe. Proposed by the French naturalist René-Antoine de Réaumur in 1730, it is based not on the properties of water but on a water-alcohol mix. The Réaumur degree (°Re) is ⅘ of a Celsius degree. Like the Celsius scale, the Réaumur scale uses the freezing point of water as 0°, which puts its boiling point at 80°Re. The conversion to Celsius is carried out using the following formulas:

$$C = \frac{5}{4} \times Re$$

$$Re = \frac{4}{5} \times C$$

6. All five temperature scales can be derived from the interval between the freezing and boiling points of water. Fahrenheit and Rankine divide this interval into 180 units; Celsius and Kelvin divide it into 100; Réaumur divides it into 80. The scales are compared in Table 4.1 and the following figure.

Table 4.1. Equivalent Temperatures

	Freezing point	*Boiling point*	*Absolute zero*
Fahrenheit	32°F	212°F	−459.67°F
Rankine	491.67°R	671.67°R	0°R
Celsius	0°C	100°C	−273.15°C
Kelvin	273.15K	373.15K	0K
Réaumur	0°Re	80°Re	

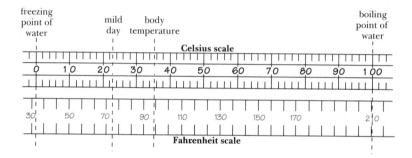

Weather and the Environment

1. Windchill Factors Because the human body gives off heat, exposed skin enjoys a thin insulating layer of body-temperature air that serves as protection from hot or cold air. But wind has the effect of removing this protection. This is why any rapid movement in a sauna seems to intensify the sensation of heat, and also why it feels colder on a windy 30° day than on a calm 30° day. The latter effect is called *windchill*. It is a perception rather than something measurable by instruments, but it is an important consideration when deciding how to dress for a cold day. A rough estimate of windchill that works for temperatures above 0°F is:

$$\text{outside temperature} - \left(\frac{3}{2} \times \text{wind speed in mph} \right) = \text{windchill temperature}$$

The National Weather Service has compiled a table that gives an idea of what various combinations of wind and cold temperatures *feel* like, reproduced below as Table 4.2. For example, on a 10° day, a 20-mph wind will cause it to feel like a −24° day with no wind. (The rule of thumb gives −20°.)

Table 4.2. Windchill Factors

Wind speed (mph) — Outdoor temperature (degrees Fahrenheit)

Wind speed (mph)	30	25	20	15	10	5	0	−5	−10	−15	−20	−25	−30	−35
0	30	25	20	15	10	5	0	−5	−10	−15	−20	−25	−30	−35
5	27	22	16	11	6	0	−5	−10	−15	−21	−26	−31	−36	−42
10	16	10	3	−3	−9	−15	−22	−27	−34	−40	−46	−52	−58	−64
15	9	2	−5	−11	−18	−25	−31	−38	−45	−51	−58	−65	−72	−78
20	4	−3	−10	−17	−24	−31	−39	−46	−53	−60	−67	−74	−81	−88
25	1	−7	−15	−22	−29	−36	−44	−51	−59	−66	−74	−81	−88	−96
30	−2	−10	−18	−25	−33	−41	−49	−56	−64	−71	−79	−86	−93	−101
35	−4	−12	−20	−27	−35	−43	−52	−58	−67	−74	−82	−89	−97	−105
40	−5	−13	−21	−29	−37	−45	−53	−60	−69	−76	−84	−92	−100	−107

2. Heat and Humidity Index Just as wind can make cold temperatures feel even colder, high levels of humidity can make high temperatures seem especially unbearable. This is because the body regulates heat through the evaporation of sweat, which causes cooling. In low humidity the air can easily absorb evaporation, whereas high humidity retards the evaporation process. Consequently a body retains more heat in humid conditions and feels hotter. Wind can provide some relief in aiding sweat evaporation, but muggy days tend to be stagnant days, with only an occasional cooling breeze.

The effect of humidity on the perception of temperature has been summarized in a table prepared by the National Oceanic and Atmospheric Administration, reproduced here as Table 4.3. Matching the temperature given in the top row to a relative-humidity reading gives the *apparent* temperature. (Relative humidity is defined in section 3 below.) Whenever this temperature exceeds 100°F, people are advised to remain indoors, or at least to minimize outdoor activities in the middle of the day.

3. Relative Humidity *Relative humidity* is a measure of how well air can absorb moisture. It is calculated as a ratio of the amount of moisture actually in the air compared to the amount of moisture the air can potentially hold under its present temperature and pressure conditions. In other words, relative humidity is the water-vapor content of the air as a percent of what it *could* be. A relative humidity of 60% means that the outdoor air is 60% saturated with moisture.

Table 4.3. Heat and Humidity Factors

Relative humidity	Perceived temperature (degrees Fahrenheit)										
	70	75	80	85	90	95	100	105	110	115	120
0%	64	69	73	78	83	87	91	95	99	103	107
10%	65	70	75	80	85	90	95	100	105	111	116
20%	66	72	77	82	87	93	99	105	112	120	130
30%	67	73	78	84	90	96	104	113	123	135	148
40%	68	74	79	86	93	101	110	123	137	151	
50%	69	75	81	88	96	107	120	135	150		
60%	70	76	82	90	100	114	132	149			
70%	70	77	85	93	106	124	144				
80%	71	78	86	97	113	136					
90%	71	79	88	102	122						
100%	72	80	91	108							

Cold air cannot hold as much moisture as hot air. Therefore, when cold outdoor air is heated to room temperature without adding any moisture, its relative humidity drops to a very low level. In addition, on cold days the moisture in a drafty house tends to be sucked out, leaving the indoor air uncomfortably dry. In houses without a vapor barrier between the insulation and the interior wall, the moisture can pass into the wall and condense, causing moisture-related problems.

When the water content of air reaches a certain level of saturation, it will condense. Thus the relative humidity helps determine when water vapor will condense into clouds, fog, or moisture on objects.

4. Dew Point and Frost Point When a cold object such as a bottle of milk is placed in warm surroundings, moisture from the air condenses on the bottle. If the bottle is warm enough, no moisture will collect on it. The exact temperature of the bottle at which moisture would begin to collect is called the *dew point.*

The *dew point* is the temperature at which an object will collect condensation (or dew). When outside temperatures are below freezing, the temperature of an object on which ice or frost will begin to form is called the *frost point.* Both the frost point and the dew point are directly related to the relative humidity of the air.

Because it represents the temperature at which moisture in the air will condense, the dew point is also the temperature of fog or a cloud formation; thus it is also known as the *cloud point.*

5. Barometric Pressure Atmospheric pressure can be seen at work whenever a suction cup is depressed. Because the air pressure is not equalized on both sides, it is the outside air pressure, which exerts a force in all directions, that keeps the suction cup on the wall. A more reliable indicator of atmospheric pressure, which operates on the same principle, is the *barometer*.

Originally, barometers were constructed as tubular glass columns filled with mercury and then inverted, so that the open end sat in a container of mercury. The weight of the mercury would create a vacuum at the top of the tube; the size of this vacuum, and thus the height of the mercury, was determined by the outside air pressing on the mercury in the container. Because barometers were originally marked off in inches, barometric pressure was also given in inches, and often still is. But for official purposes, other units of measure are used.

Air pressure can vary in response to changes in temperature and humidity. But the primary determinant is altitude. With greater altitude, the air becomes thinner and thus exerts less pressure. One *atmosphere* is defined to be the air pressure at sea level. This corresponds to about 30 inches, or 760 millimeters, of mercury. In the metric system, the base unit of pressure is the *pascal*. Atmospheric pressure can be expressed in pascals; in *millibars*, units of 100 pascals; in *kilopascals*, units of 1000 pascals; or in *bars*, units of 100 kilopascals. Although the standard unit of measurement for barometric pressure is the kilopascal, weather reporters still like to refer to *inches* of mercury. Conversion factors are given in Table 4.4.

Table 4.4. Conversion of Air-Pressure Units

1 kilopascal = 10 millibars = .295 inches of mercury

1 inch of mercury = 3.39 kilopascals

1 atmosphere = 760 mm of mercury = 29.92 inches of mercury

1 atmosphere = 101,325 pascals = 1013.25 millibars = 1.01325 bar

1 millibar = 0.75 mm of mercury ≈ $\frac{1}{32}$ inch of mercury

1 inch of mercury = 0.03342 atmospheres

Zones of pressure on weather maps are indicated by *isobars*. Each isobar represents a region of uniform pressure. In general, low-pressure weather systems are associated with stormy weather, and high-pressure systems with clear, calm weather.

6. The Beaufort Scale In 1805 the English naval commander (and later admiral) Sir Francis Beaufort devised a scale to measure wind velocity at sea. He based the scale on the sail requirements of a "well-conditioned man of war" in different wind conditions. The scale runs from 0 to 12.

Beaufort's scale was widely adopted for maritime use by the mid-1800s. Several attempts were made to place it on a mathematical basis—to have it reflect sea disturbance and even the effects of wind on land. The scale is still used by meteorologists. It is given in Table 4.5 with the official wind-name classifications of the U.S. National Weather Service.

Table 4.5. The Beaufort Scale

Beaufort number	Name of wind	Speed at a height of 10 meters		
		Knots	Miles/hr	Km/hr
0	Calm	under 1	under 1	under 2
1	Light air	1–3	1–3	2–5
2	Light breeze	4–6	4–7	6–11
3	Gentle breeze	7–10	8–12	12–19
4	Moderate breeze	11–16	13–18	20–29
5	Fresh breeze	17–21	19–24	30–39
6	Strong breeze	22–27	25–31	40–50
7	Moderate gale	28–33	32–38	51–61
8	Fresh gale	34–40	39–46	62–74
9	Strong gale	41–47	47–54	75–87
10	Whole gale	48–55	55–65	88–101
11	Storm	56–65	65–75	102–120
12	Hurricane	over 65	over 75	over 120

Time

Time is a way to specify either an instant on a continuum or an interval of duration. Units of time have always been based on natural phenomena—on the motions of the sun, the moon, and the stars. But as technology has made greater demands for precision in time measurement, the irregularity of such periodic phenomena has led to a search for better ways to keep time.

The divisions of time we currently use come down to us from the ancient Sumerians and Egyptians, who passed their 60-second minute and 60-minute hour on to the Babylonians and Greeks. The hour as a fixed unit has not always meant what we mean by it. The Egyptians, for example, divided daylight into 12 equal intervals, and night into 10 equal intervals, designating the remaining two intervals as dusk. This system—called *apparent solar time*—is agriculturally motivated, and it survived well into the Middle Ages. It made sense because it divided the workday into equal units, making the sun (or a sundial) the standard timekeeper. Although such hours are longer in the summer than in the fall, the system had the advantage of letting the farmer know what portion of the working day remained.

The idea of dividing the day into 24 hours goes back to the ancients, and although it did occur to the Egyptians to experiment with a decimal system, not until the reforms of the French Revolution would a true decimal time system be proposed on a grand scale. It consisted of a 10-day week, a 10-hour day, a 100-minute hour, and a 100-second minute. And it lasted only about 10 years, from 1795 to 1805. The French people could accept a decimal clock, but they drew the line at a 10-day week, which ultimately did the system in.

The modern metric system defines the second as the basic unit of time, and scientific time measurement is all based on the second.

Table 4.6 lists a number of scientific units of time measurement, most of them based on the second.

Table 4.6. Units of Time

femtosecond (fs)	10^{-15} sec	quadrillionth of a second
picosecond (ps)	10^{-12} sec	trillionth of a second
nanosecond (ns)	10^{-9} sec	billionth of a second
microsecond (μs)	10^{-6} sec	millionth of a second
millisecond (ms)	10^{-3} sec	thousandth of a second
one second		
kilosecond (Ks)	10^{3} sec	thousand seconds ≈ 16.5 min.
megasecond (Ms)	10^{6} sec	million seconds ≈ 11.5 days
gigasecond (Gs)	10^{9} sec	billion seconds ≈ 32 years
terasecond (Ts)	10^{12} sec	trillion seconds ≈ 32,000 years

1 mean solar day = 1.0027379093 sidereal days

1 sidereal day = 0.9972695664 mean solar days

1 solar year = 365.242194 solar days, or 365 days, 5 hours, 48 minutes, 45.51 seconds

1 sidereal year = 366.2564 sidereal days = 365.25636 mean solar days

If one year is defined as the basic unit of time, then months, weeks, days, hours, minutes, and seconds can all be defined in relation to it. But just as there are many kinds of days, there are many kinds of years, all of slightly different lengths. There is the calendar year, the lunar year, the sidereal year, the solar year, and the fiscal year. Consequently there are calendar days, sidereal days, lunar months, sidereal months, and so on. The methods of determining lengths of time are briefly described below.

1. *Solar time* refers to the motion and elevation of the sun, or more properly to the earth's motion with respect to the sun. A *solar year* or *tropical year,* for example, is the average duration of the earth's orbit around the sun as measured by the sun's *ecliptic*—its apparent path across the sky. A solar year, as specified above, is 365 days, 5 hours, 48 minutes, and 45.51 seconds (or 365.2422 days).

2. *Sidereal time* refers to timekeeping based on the movement of the stars. A sidereal year is the period in which the Earth completes a full circuit around the sun, as it would appear to a viewer on a distant star. Because the stars are the most consistent long-term timekeepers, astronomers and other scientists use them as the basis for scientific clocks.

A sidereal year lasts 366 sidereal days, 6 hours, 9 minutes, 9.54 seconds. What at first seems to be an important discrepancy between the solar year and the sidereal year is easily reconciled. The figure given above is based on the *sidereal day,* which, because it is derived from the sidereal year, is about four minutes shorter than the *solar day.* When the sidereal year is measured in solar days, it turns out to be only about 20 minutes longer than the solar year.

During the course of a year, a given star appears to revolve around the earth one more time than the sun does. This is because during the course of the year, the earth makes one complete revolution around the sun. Thus a star-determined day (a sidereal day) is about 3 minutes, 55.91 seconds shorter than a mean solar day. This difference, over 365.25 (mean solar) days, makes up for the one-day difference between a sidereal year and a solar year.

3. *Lunar time* can be measured in two ways: with respect to the stars or with respect to the sun.

A *sidereal lunar month* is the time it takes the moon to complete one orbit around the earth as it would appear to a viewer on a distant star— about 27⅓ days. This is the same as a *lunar day* because it coincides with a single rotation of the moon about its axis. (Recall that the same side of the moon always faces the earth.)

Synodical time offers another way of calculating a lunar cycle. A *synodical month,* also called a *lunar month,* is the period from one new moon to the next, which takes about 29½ days. Physically, this is the time between two successive alignments of the moon between the earth and the sun.

CALENDARS

The word *calendar* comes from the Latin *kalends,* meaning the day on which accounts were due. In its modern sense, a calendar is a convenient division of a year into months, weeks, and days. But the urgency of its Latin origin remains. The calendar is a necessary means of scheduling meetings, assigning deadlines, and setting due dates, and its consistency has always been an important issue.

1. The Egyptians first conceived of a year as lasting 360 days, basing their calculations on the flooding schedule of the Nile. Later, when the sun's deviation from this schedule became apparent, five days were added. More refinement was required, however, so that solar events such as the solstices and equinoxes would continue to occur on roughly the same dates every year. The Egyptians eventually came to the conclusion that the year consisted of 365¼ days.

2. Julius Caesar instituted the *Julian calendar* in the year 45 B.C., originating the idea of a *leap year*—one year in every four in which there is an extra day. In Caesar's era the new year began on March 1; the leap day was added at the end of the Roman year, which is how it came to fall on February 29.

The Romans accepted the Egyptians' calculation of a 365¼-day solar year, but in reality the solar year is about 11 minutes shorter than this. As a result, the Julian calendar would gain a full day every 120 years, and this would cause calendar dates to fall behind corresponding solar events. By the year 1000, for example, the spring equinox was occurring eight days short of March 21. Because this time lag made it nearly impossible for the Church to schedule its movable feasts, a revision of the Julian calendar became necessary.

3. In 1582 Pope Gregory XIII ordered that 10 days be dropped from the calendar so that the spring equinox would fall on March 21. And to correct the error in the Julian calendar, the Pope's decree eliminated three leap years in every 400 (specifically, on those century years not divisible by 400). Because it works, the *Gregorian calendar* has become the worldwide standard, although it took several centuries to take hold globally, and was not adopted in Britain and the American colonies, for example, until 1752, or in Russia until 1918.

4. A *lunar cycle* is the time required for the moon to complete one set of *phases*. It begins with a *new moon*, which occurs when the moon rises at the same time as the sun and thus reflects no sunlight. As the cycle progresses, the moon *waxes* (or increases in size and visibility) each night until it reaches its full stage two weeks later. It then *wanes*, or gradually decreases back to crescent stage and to the next new moon. A full moon rises at sunset, and each night thereafter rises later and later as it wanes. The new moon rises at dawn.

Φ —————————————————————————————— Φ

THE YEAR ZERO

The way we number years follows a suggestion made by the Roman abbot Dionysius Exiguus about 525 A.D.. His proposal contained a number of errors and an error of numbers. In attempting to refer all events to the birth of Christ, Dionysius placed that date several years too late. Even more inconveniently, he left out the year 0.

On a number line, the interval between any two numbers is equal to their difference. This is not true, however, on the timeline of Dionysius. Due to an error of continuity, the progression A.D. 3, A.D. 2, A.D. 1 is not followed by the year 0, but by 1 B.C., which is numerically equivalent to -1. Thus the difference between A.D. 1 and 1 B.C. is not two years (which is the difference between 1 and -1) but one year. The period from 20 B.C. to A.D. 10 might at first glance seem to be 30 years, but is in fact only 29. Astronomers sidestep this by using positive and negative signs instead of A.D. and B.C. Thus:

1998 (historical) = +1998 (astronomical)

1 B.C. (historical) = 0 (astronomical)

200 B.C. (historical) = -199 (astronomical)

Identifying the days within a given historical year is complicated by irregularities in calendar systems. As a result, astronomers have adopted a simple system based on a successive numbering of days. These *Julian days* are described below in section 8.

Φ —————————————————————————————— Φ

5. The lunar cycle, or lunar month, lasts about 29½ days, so 12 lunar cycles last about 354 days, which is 11¼ days short of a full year as measured by one earth revolution about the sun. Consequently the moon is not a suitable long-term timekeeper.

Because the lunar month is shorter than most calendar months, there can be two full moons in one calendar month. The second full moon is called a *blue moon*. If a full moon falls on the first or second day of a 31-day month, there will be a blue moon at the end of the month, but this occurs infrequently, which gives rise to the expression "once in a blue moon."

6. Calendars acknowledge the lunar cycle by dividing the year into 12 units. But each month does not begin with a new moon. This is because a full year is determined by the movement of the sun, or, more accurately, by the movement of the earth around the sun.

A *solar year* is the duration of one earth orbit around the sun. Within the solar year are four important days—the two *equinoxes* and the two *solstices*. The summer and winter solstices are the longest and shortest days of the year in terms of amount of daylight in the northern hemisphere. The equinoxes (the word means literally "equal night") fall midway between the solstices and are the days when day and night are of equal length. These four days divide the year into the four seasons.

> Spring (or vernal) equinox: March 21
> Summer solstice: June 21
> Autumnal equinox: September 21
> Winter solstice: December 21

(These divisions may also fall on the 20th, 22nd, or 23rd of the month.)

7. The length of a week and a month have little to do with any natural phenomena. A month is close to a lunar cycle, but the number of days allotted to each month—from 28 to 31—was a matter of arbitrary decision and decree by the Roman senate.

8. Julian Days *Julian days* form the basis of a calendar that uses a consecutive numbering of days. This calendar commences with Julian Day One on noon of January 1, 4713 B.C. (designated January 1.0, −4712, in astronomical time). The reasoning behind this choice turns out to have been faulty, but the advantages of a continuous calendar have little to do with the day on which it begins.

The Julian day is the invention of the Italian scholar Joseph Scaliger (1540–1609). Although its name makes for easy confusion with the Julian calendar, it was inspired not by Caesar but by Scaliger's father, Julius. The rationale behind the choice for the first day depends upon three cycles deemed important (or useful) by Scaliger:

> i. In any year a particular date falls on the same day of the week that it did 28 years ago. This is called a *solar cycle,* and is the basis of perpetual calendars.
>
> ii. The phases of the moon always fall on the same day of the week that they did 19 years earlier. This is referred to as the *lunar cycle.*
>
> iii. Under the emperor Diocletian (reigned A.D. 284–305), the tax census occurred every 15 years, a period referred to as the *cycle*

of indiction, which was chosen by Scaliger principally for the sake of the number 15.

Combining the three cycles yields a period of 28 × 19 × 15 = 7980 years of days in which no three cycles coincide. Scaliger named this period of 7980 unique days a *Julian period.* He undoubtedly chose the cycle of indiction because it extended the product of the lunar and solar cycles into a period large enough to contain all of recorded history. In fact his choice of 4713 B.C., which he calculated to be the last time the first day of all three cycles coincided, conveniently coincided with what some medieval Biblical scholars proposed as the date of the Creation.

Scaliger miscalculated, but no matter. He argued that his calendar provided a useful way of keeping records of astronomical observations, and it has survived as just that. In order to avoid the confusion caused by different calendar systems, leap years, and eliminated days (such as the ten days excised by Pope Gregory), astronomers refer to Julian days. They divide the day into tenths rather than hours. The stroke of midnight ushering in January 1 in the year 2000 falls on Julian day 2,451,179.5. (The decimal .5 results from the fact that Julian days begin at noon.)

The so-called *modified Julian date* is an adjustment to the Julian system that makes the numbers easier to deal with and also makes the day start at midnight instead of noon. It is calculated by subtracting 2,400,000.5 from the Julian day, which places Day One on 17 November, 1858. This system appeals primarily to computer programmers, who have their own problems with traditional calendars.

Geography, Navigation, and Travel ───────────

MAPS

A *map* is a way of representing space. The most familiar maps represent the whole or part of the earth's surface. But there are also maps that represent the moon, the surfaces of other planets, and the heavens, as well as interior spaces (such as building directories) and prevailing conditions (such as weather maps, population density maps, or other statistical maps). This section will focus on the mathematical problem of representing the spherical surface of the earth on the flat surface of a map.

1. For purposes of mapmaking, the earth is modeled as a perfect sphere that rotates on a *polar axis* and is encircled by an *equator.*

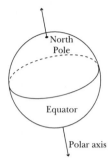

2. Any place or point on the globe can be specified using two numbers. The first denotes the point's angular elevation above or below the equator. This is called *latitude*. Zones of latitude are shown on a globe by circles that run parallel to the equator.

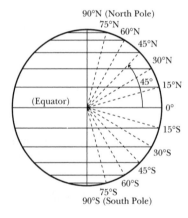

LATITUDES

Any circle of latitude can be identified by its angle of elevation above the horizontal from the earth's center. The north pole is at ninety degrees north latitude, or 90°N. The south pole is at 90°S. New York City is at about 40°45′N.

In geographical measures, 1 degree is divided into 60 minutes, and each minute into 60 seconds. At the equator, a degree of latitude is equivalent to about 110 km, a minute to 1.8 km, and a second to about 31 meters. (U.S. equivalents are 69 miles, 1.15 miles, and 34 yards.) These distances vary slightly at latitudes farther to the north or south because the earth is not a perfect sphere.

3. The second number used to locate a point on the earth's surface denotes how far east or west the point lies from Greenwich, England. This measure is called *longitude*. The line that runs directly north and south through the Greenwich Observatory ultimately encircles the globe, passing through both poles. The semicircle running from pole to pole through Greenwich is called the *prime meridian*. In general, a *meridian* is any semicircle that connects the poles. Every point on the globe lies on a meridian, which places it at a certain angle east or west of the prime meridian.

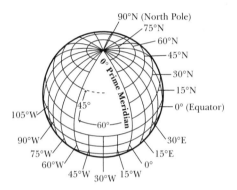

The longitude scale runs from 0° to 180° East, and 0° to 180° West, the two designations meeting at the opposite side of the globe from Greenwich, at the International Date Line. New York City has a longitude of 74°W; Moscow has a longitude of 37°30′E.

One degree of longitude at the equator is equivalent to about 110 km or 69 miles. This diminishes steadily as longitude lines converge to the north and south. Table 4.7 on page 172 shows the convergences of distances representing 1° of longitude as the latitude increases.

4. The earth is not a perfect sphere, although it is convenient to think of it that way. During the long period of its formation, the earth's rotation caused some widening at the equator and some flattening out at the poles. But the distortion is not great. The earth's polar radius (the distance from its center to either pole) is 6357 kilometers, whereas its equatorial radius (the distance from the center to the equator) is 6378 kilometers. The difference amounts to less than one tenth of a percent of either radius.

5. To understand latitude and longitude and their relationship to distances along the earth's surface, it is useful to know something about the geometry of a sphere. This in turn requires some knowledge of circles. Lines of longitude form circles known as great circles because they are the largest circles that can be drawn on the surface of a sphere. A great circle is any circle that passes through two opposite points—essentially, points that lie at either end of a diameter of a sphere. In the case of longitude lines, the two points are the north and south poles. Lines of latitude also form circles, but only one of these—the equator—is a great circle.

6. The shortest distance between any two points on the surface of a sphere is along the path of a great circle. Such distances can be calculated using latitude and longitude coordinates and from certain facts relating to *circumference.*

The circumference of a circle is the distance around it. The formula for the circumference is based on the radius or the diameter (which is twice the radius).

$$\text{circumference} = 2 \times \pi \times \text{radius} = \pi \times \text{diameter}$$

Any angle drawn from the center of the circle will mark off part of the circle. This part is called an *arc* of the circle. Its length is proportional to the circumference according to this formula:

$$\frac{\text{length of arc}}{\text{circumference}} = \frac{\text{degree measure of angle}}{360}$$

which is equivalent to:

$$\text{length of arc} = \frac{\text{degree measure of angle}}{360} \times \text{circumference}$$

In the figure below, a circle of radius 10 has a circumference of 20π units. An angle of 60° marks off an arc whose length is $^{60}/_{360}$ times the circumference. This amounts to $20\pi \times {}^{60}/_{360} = 20\pi \times \frac{1}{6} = {}^{20\pi}/_{6} = 10.47$ units. (Like all measurements involving π, this result is rounded off.)

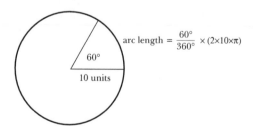

$$\text{arc length} = \frac{60°}{360°} \times (2 \times 10 \times \pi)$$

As will be shown on page 171, if the angle between two points on a great circle is known (and this can be found using latitudes), the distance between the points can be found by the process outlined above, where the radius of the earth is used as the radius of the circle.

Φ ——————————————————————————————— Φ

TIME ZONES

Time zones originated in the United States in 1883 at the request of railroad executives who needed a standard time by which to coordinate their train schedules. From this were born the original four U.S. time zones—Eastern, Central, Mountain, and Pacific. Passing westward into a new time zone moves the clock back one hour; therefore clocks in the Pacific time zone are set three hours earlier than those in the Eastern time zone. Time zones extend both east and west around the globe, so that Paris is six hours ahead of New York and Hawaii is six hours behind New York.

Because this procession of time zones has to end somewhere, an *International Date Line* was arbitrarily established at about the 180° meridian (opposite Greenwich, England). The date line does not follow the meridian precisely; it actually zigzags in order to avoid land masses and archipelagoes. Passing over this line from east to west results in moving forward a calendar day; from west to east takes one day back. (This is how Jules Verne's Phileas Fogg won his bet in *Around the World in Eighty Days*.)

The difference between time zones is further complicated by *daylight saving time*. This was instituted early in this century to adjust for the lengthening of daytime in summer months. The change from standard time to daylight saving time occurs in the United States at 2:00 A.M. on the first Sunday in April, and standard time returns at 2:00 A.M. on the last Sunday in October.

The changing of the clocks is not observed everywhere in the United States; and although other countries also have daylight saving time, there is no universal agreement as to when it should begin and end. (European countries use the last Sunday in March and the last Sunday in September.)

Φ ——————————————————————————————— Φ

7. The idea of a map is inseparable from the idea of scale. A *scale* is the size of the representation of an object relative to its actual size. A map might use a centimeter to represent one kilometer, for example. The scale in this case would be 1 centimeter to 100,000 centimeters. This is written as $\frac{1}{100000}$, or as 1 : 100,000. A map on which 1 inch represents 10 miles has a scale that can be written as "1 in. to 10 mi." or as "1 : 633,600" (because there are 633,600 inches in 10 miles).

In general, a scale is a ratio of the number 1 to some larger number; the larger the second number, the smaller the scale of the map. (A 1 : 1 scale would be actual size.) A trade-off arises between scale and precision. A smaller scale allows a greater area to be represented. A larger scale allows greater detail to be shown.

U. S. Geological Survey maps are produced in several scales, from detailed local maps (1 : 24,000) to more inclusive regional maps (1 : 150,000). These maps come in two forms: planimetric and topographic.

A *planimetric map* includes outlines of those physical features, natural and manmade, that the scale permits. These may include rivers, streams, roads, houses and buildings, airports, and so on.

A *topographic map* has the same features, but also includes elevations denoted by contour lines. A *contour line* indicates a path of constant elevation above sea level. The simplest and most visible example of a contour line is provided by the boundary of a lake or pond. On maps, contour lines are drawn at intervals of 10 feet or 100 feet, depending on the scale. The spacing of the line is an indication of the steepness of the terrain: the closer the lines, the steeper the slope.

8. *Statistical maps,* like topographic maps, associate a numerical quantity (such as population, mean income, incidence of disease) with a geographical region. This can be done by color coding, by using dots to represent some numerical count, or by sizing the region to correlate with its percent representation. In the population distribution map shown in the figure on the following page, the area of each state has been scaled to reflect its share of the national population (1970). Note how Idaho appears much smaller than it really is, and New Jersey much larger.

9. World Maps and Projections The problem of representing the surface of the entire globe on a flat map while preserving all distances and angles cannot be solved. In any scheme, there will either be distortion of area or distortion of the distances between points. One map that distorts both area and distance but that preserves compass settings and angles between points is the map preferred by navigators; it is called a *Mercator projection.*

The Mercator projection is named for Gerardus Mercator (1512–94), the Flemish mathematician and cartographer who invented it. His idea can be visualized by imagining a transparent globe with a light source at its center. If a rectangle of paper is rolled into a tube and

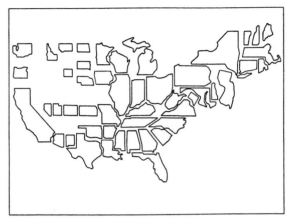

Source: Phillip C. and Juliana O. Muehrcke, *Map Use: Reading, Analysis, and Interpretation*, 3rd ed. (Madison, Wisc.: JP Publications, 1992). Reprinted by permission.

placed around the globe so that it touches the equator all the way around, the outlines of the continents would be projected onto the inner surface of the tube. Points near the equator would suffer little distortion, but land masses farther to the north and south would become exaggerated in size. Lines of latitude would become horizontal, longitude lines would be projected as vertical lines, and together they would form a rectangular grid. See the figure on page 170.

The cylindrical projection, however, while similar to Mercator's solution, does not preserve compass readings perfectly. Mercator solved this problem not with a projection (the term *Mercator projection* is therefore not quite correct), but by stretching intervals of latitude in the same proportion as lines of longitude, which are bent into straight lines to form the verticals on a rectangular grid. In Mercator's words: "In making this representation of the world we have had to employ a new proportion and a new arrangement of the meridian [of longitude] with reference to the parallels [of latitude]. . . . We have progressively increased the degrees of latitude towards each pole in proportion to the lengthening of the parallels with reference to the equator."

The calculation on a perfect sphere is a simple one, but because the Earth is not quite a perfect sphere, more sophisticated methods must be used to produce accurate maps. Still, here is the gist of Mercator's method:

On the surface of a globe lines of latitude form circles parallel to the equator, with circumferences smaller than the equator. Lines of longitude run through the poles. The ratio of the circumference of a circle of latitude to the circumference of the equator represents the amount of compression in the vertical lines of longitude. In making latitude and longitude lines into a grid, Mercator widened the intervals between successive latitude (horizontal) lines by the same ratio that longitude lines are "stretched" at those latitudes. Thus the latitude lines (usually shown in 10° increments) grow farther apart toward the poles. This results in a great distortion of area near the top and bottom of the map, with the least distortion near the equator. At any point on the map, however, due north is straight up, and a line connecting any two points on the map corresponds to the true compass bearing relative to north.

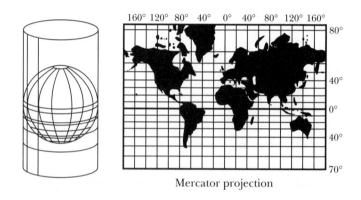

Mercator projection

DISTANCE, LATITUDE, AND LONGITUDE

1. To find the difference in latitude between two cities that lie on the same meridian in the same hemisphere (north or south), convert each latitude to decimal form and subtract the smaller from the larger latitude.

- St. Louis and New Orleans lie on the 90°W meridian. Their latitudes are 38°40′N and 30°0′N, respectively. The difference is 8°40′, which is 8.67°. (To convert 40′ to decimal form, write it as 40/60 and convert to its decimal equivalent, 0.66, which can be rounded to 0.67.)

- Miami and Pittsburgh lie on the 80°W meridian. Their latitudes are 25°45′N and 40°30′N, respectively. To find the difference in latitude, borrow 60 minutes from Pittsburgh's 40° of latitude, thus converting its coordinate to the equivalent latitude of 39°90′. Now subtract 25°45′ to get 14°45′, or 14.75°.

2. To find the difference in latitude between cities that lie in opposite hemispheres, add their latitudes.

• Johannesburg and St. Petersburg are both approximately 30° east of the prime meridian. Johannesburg is at 26°10′S latitude, and St. Petersburg is at 60°N. The difference in latitude is 26°10′ + 60° = 86°10′, meaning 86¹⁰⁄₆₀°, or 86.167°.

• Montreal and Santiago, Chile, are close to the same meridian. Montreal is at 45°30′N latitude, and Santiago is at 32°40′S latitude. Their difference in latitude is 45°30′ + 32°40′ = 78°10′, or 78.167°.

3. To find the distance between two cities on the same meridian:

   i. Find the angular distance in latitude between the two cities in decimal form.

   ii. Multiply the angular distance by 69 (for miles) or 111 (for kilometers).

   • The distance between St. Louis and New Orleans, with a difference in latitude of 8.67°, is 8.67 × 69 ≈ 600 miles.

   • The distance between Miami and Pittsburgh, with a difference in latitude of 14.75°, is 14.75 × 69 ≈ 1020 miles.

The multipliers come from the formula for the circumference of a circle, which is π times the diameter. The ratio of the angular difference between cities to the number of degrees in a full circle (360°) should be the same as the ratio of the distance between cities to the circumference of the earth, where

circumference of earth = π × 12,750 km ≈ 40,000 km

= π × 7920 miles ≈ 24,900 miles

This results in the proportion:

$$\frac{\text{distance between cities}}{\text{circumference of Earth}} = \frac{\text{angular difference}}{360°}$$

Or, in equivalent form,

$$\text{distance between cities} = \text{angular difference} \times \frac{\text{circumference of Earth}}{360°}$$

   • The distance between St. Petersburg and Johannesburg, with an angular difference of 86.167°, is about 86 × 111 ≈ 9500 kilometers.

   • The distance between Montreal and Santiago, with an angular difference of 78.167°, is about 78 × 111 ≈ 8700 kilometers.

4. It is also possible to compute the east-west distances between cities, but the formula is more complicated because the distance-equivalent of one degree of latitude decreases as you move from the equator to the poles. Thus the multiplier changes. At the equator, a difference of one degree of latitude or longitude represents the same distance: about 111 kilometers, or 69 miles. At 50° north latitude, however, the east-west distance-equivalent to a one-degree difference of longitude is about 72 kilometers, or 45 miles.

Table 4.7 gives appropriate multipliers for different latitudes, which are equally valid in the Northern and the Southern hemisphere. For example, New York and Hamburg, Germany, are both close to the 40th parallel of northern latitude. Their difference in longitude is about 84°. The multiplier that converts angular longitudinal difference to kilometers at the 40th parallel is approximately 85. Thus the east-west distance between New York and Hamburg is approximately 85 × 84, or 7140 kilometers.

- Stockholm, Sweden, at 59°23′N latitude and 18°00′E longitude, and Anchorage, Alaska, at 61°12′N latitude and 149°48′W longitude, are both close to the 60th parallel. Their difference in longitude is found by summing their longitudinal angles (because one is east and the other west). By first rounding Anchorage's reading to 150°W, the total comes to 150 + 18 = 168°. According to Table 4.7, the multiplier for 60°N is about 56 for kilometers and 35 for miles. Thus Stockholm and Anchorage are approximately 168 × 56 ≈ 9400 kilometers, or 168 × 35 ≈ 5900 miles apart along the 60th parallel.

Table 4.7. Distance Equivalence of 1° of Longitude at Various Latitudes

At a latitude of	1° of longitude equals		
	km	miles	naut. miles
0	111.32	69.17	60.11
10	109.64	68.13	59.20
20	104.65	65.03	56.51
30	96.48	59.95	52.10
40	85.39	53.06	46.11
50	71.70	44.55	38.71
60	55.80	34.67	30.13
70	38.19	23.73	20.62
80	19.39	12.05	10.47
90	0.00	0.00	0.00

Φ ——————————————————————— Φ

DISTANCE TO THE HORIZON

The horizon, as hazy and indistinct as it is, exudes a certain air of the exotic. This is partly because it is not entirely clear how far away it is. A sailor sits in the crow's nest, scanning the horizon for any sight of land or ships, or perhaps even a great whale. But how far off can he see?

A rough calculation of the distance to the horizon is provided by a simple formula based upon classical geometry. It says:

$$\text{distance to horizon (in miles)} \approx \sqrt{\frac{3}{2} \times \text{height of vantage point (in feet)}}$$

Thus a six-foot man standing at the ocean's edge sees approximately 3 miles out. A sailor looking out to sea from a point 30 feet up a mast sees $\sqrt{45}$ or almost 7 miles in any direction.

The ratio ³⁄₂ comes from the two special numbers—the diameter of the earth, which is about 7920 miles, and the number of feet in a mile, which is 5280. The proof is too long to go into here, but the figure below shows how the problem is set up and gives the proportions that lead to the formula. The rest is left as an exercise.

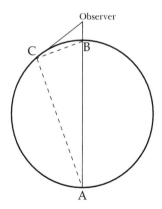

In the figure, triangle BOC is similar to triangle COA. Thus $^{OC}/_{OA} = {}^{OB}/_{OC}$. By cross-multiplication, $OC^2 = OA \times OB$. Because BA is the diameter of the earth, it can be substituted for OA (since OB, the height of the observer, is relatively small). Thus $OC^2 = BA \times OB$. The rule of thumb for distance to the horizon results when BA is expressed in miles (about 7900 miles) and the height of the observer, OB, is also expressed in miles (divide height in feet by 5280). $7900/5280$ is approximately $3/2$. The last step is to take the square root.

Φ ———————————————————————— Φ

BEARINGS AND AZIMUTHS

1. A *bearing* or an *azimuth* gives a direction in the form of a degree angle measured from a given reference direction. Bearings are measured from the north or the south, clockwise or counterclockwise. Thus a bearing never exceeds 90°. Azimuths are clockwise angles, usually measured from a meridian line, so azimuths use either north or south as their reference direction.

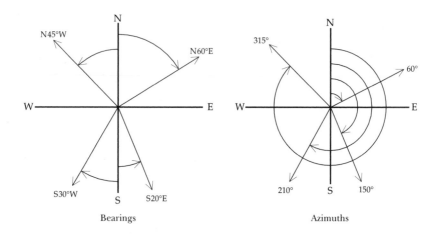

Bearings   Azimuths

In this figure, note how a bearing specifies a quadrant, whereas an azimuth does not. For an azimuth based on true north:

0° to 90° covers the northeast quadrant.
90° to 180° covers the southeast quadrant.
180° to 270° covers the southwest quadrant.
270° to 360° covers the northwest quadrant.

2. Astronomers use azimuths in combination with altitude to describe the position of objects in the skies. The *altitude* is the angle of elevation above the horizon (up to 90°). The azimuth is the number of degrees measured clockwise from due south (up to 360°). For non-astronomical purposes such as navigation or artillery sighting, the azimuth is measured from due north.

3. Bearing and range may be converted to rectangular coordinates (see page 67). To translate azimuth and range into rectangular coordinates, use these conversion formulas:

  i. The north-south distance is given by the formula: *range × cosine of azimuth*. If the result is positive, the direction is north; if negative, south.

 ii. The east-west distance is given by the formula: *range × sine of azimuth*. If the result is positive, the direction is east; if negative, west.

4. Calculating the azimuth and range is a standard means of aiming artillery or directing anything that travels "as the crow flies." All that is required is a direction and a distance. In situations where a direct route is impossible and the shortest distance between two points is not a straight line, it may be necessary to use what mathematicians call the "taxicab metric." To get from one point to another in almost any major American city involves traveling on a grid, where the only possible directions of travel are east-west and north-south. These represent *rectangular coordinates,* and in such situations, azimuth and range are relatively useless.

 ♦ An azimuth of 300° and a range of 3000 meters converts to rectangular coordinates as:

$$\text{range} \times \text{cosine of azimuth} = 3000 \times \cos 300°$$

$$= 3000 \times (0.5)$$

$$= 1500 \text{ meters}$$

The positive value indicates north.

$$\text{range} \times \text{sine of azimuth} = 3000 \times \sin 300°$$

$$= 3000 \times (-0.866)$$

$$\approx -2600 \text{ meters}$$

The negative value indicates west. (Of course, in a true *taxicab* metric, the proper unit of measure would be *blocks*.)

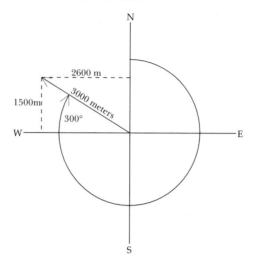

Astronomical Distances

The measurement of distances between the planets or to the stars requires some special techniques involving trigonometry. The reader might find it useful to review the relationships between the sides and angles of right triangles discussed on pages 103–108. Measuring astronomical distances is a natural application of trigonometry, because angles between visible objects are easier to measure than the distances themselves. For example, if you stand at the point labeled A in the figure below, and walk to point B or C, your line of sight to point P changes. The angle measure of this change depends on how far you walk, and in what direction. For the sake of simplicity, let's assume you walk on a path that is perpendicular to the line AP. The change in your line of sight is referred to as the *parallax* of point P. The distance you walk, either from A to B or A to C, is referred to as the *baseline distance*.

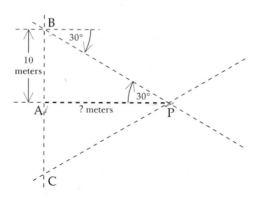

Suppose a sighting device gives you the measure of the parallax as 30°, as shown in the figure, and the baseline distance is 10 meters. From this information you can find the distance from A to P by using the following formula:

$$\tan 30° = \frac{\text{opposite side}}{\text{adjacent side}} = \frac{AB}{AP} = \frac{10 \text{ meters}}{AP}$$

This means that the tangent of 30° is equal to 10 divided by the distance AP. Because we want to know the distance AP, we *solve for* AP using the algebraic techniques of Chapter 2, and get

$$AP = \frac{10}{\tan 30°}$$

A calculator or table of values gives the tangent of 30° as about 0.577. Thus the distance from A to P is (10 ÷ 0.577), or about 17.33 meters.

DISTANCE FROM EARTH TO SUN

1. The trigonometric formula: *tangent equals opposite side over adjacent side* can be used to find astronomical distances—the distance from the earth to the moon, from the earth to the sun, and from the sun to the stars.

2. The distance from the earth to the sun can be calculated from a known distance and a known angle. The largest base distance ready to hand is the radius of the earth, which is about 6378 km at the equator. The angle can be measured by taking advantage of the earth's rotation.

3. Assume you are at point A viewing a point on the sun (a sunspot, for example) through a telescope. Twelve hours later you are at point B, 12,756 km from your previous vantage point. Because of the change in position, you have to adjust the telescope setting slightly. The angle through which the telescope must be adjusted will be the same as the angle at S in the figure below. Half of this angle, labeled *x*, is called the *parallax of the sun.*

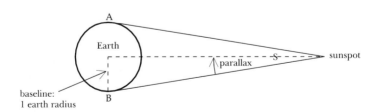

The telescope would have to be adjusted by 17.6 seconds of arc. (One second is ¹⁄₆₀ of a minute, which is ¹⁄₆₀ of a degree, so 1 second is

$\frac{1}{3600}$ of a degree.) Thus the parallax of the sun would be half of the measured angle, or about 8.8 seconds of arc (which is equivalent to $8.8/3600 = 0.002444$ degrees).

The tangent formula:

$$\text{tangent of parallax} = \frac{\text{radius of the earth}}{\text{distance to the sun}}$$

can be rewritten as:

$$\text{distance to the Sun} = \frac{\text{radius of the earth}}{\text{tangent of parallax}}$$

$$= \frac{6378 \text{ km}}{\text{tangent of 8.8 seconds}}$$

$$\approx 149,500,000 \text{ km}$$

The figure is not exact because the method of measuring the angle has been overly simplified. Some adjustments and corrections were ignored. During the 12 hours between measurements, for example, the earth moved some distance along its orbit around the sun. Because the earth does not afford a fixed observation platform, background objects such as very distant stars may be used to align telescopes in order to measure parallax more accurately. (The true parallax of the sun is closer to 8.798 seconds of arc.) Still, the result given above is within 0.06% of the accepted value of 149,597,870 km.

4. The mean distance from the earth to the sun is defined as one *astronomical unit* (AU). Astronomical units are commonly used to express distances within the solar system. The distance from the earth to the moon, for example, is about 0.0027 AU. The distance from the sun to Jupiter is 5.2 AU. See Table 4.8 for other planetary distances.

Table 4.8. Planetary Distances from the Sun

Planet	Millions of kilometers	Millions of miles	Astronomical units
Mercury	58	36	0.387
Venus	108	67	0.723
Earth	149	93	1
Mars	228	142	1.52
Jupiter	779	484	5.20
Saturn	1428	887	9.54
Uranus	2872	1784	19.18
Neptune	4500	2794	30.06
Pluto	5900	3666	39.44

DISTANCES TO STARS

Parallax is also used to estimate distances to stars, and it serves as the basis for another unit of astronomical measure: the *parsec*.

1. The earth's diameter is large enough to serve as the base of a triangle for calculating the distance to the sun. But because the distance even to the nearest star is so much greater than the earth's distance to the sun, a baseline distance larger than the earth's radius must be used. This is provided by the *astronomical unit*, which is the earth's mean radius of revolution about the sun. If a star is sighted on January 1st, and again on July 1st, by which time the earth has completed a half-orbit, it will be 2 AU removed from its former position. Even with this base, the largest parallax that can be measured is about three-fourths of a second of arc. This is the parallax of Alpha Centauri, the nearest star cluster.

2. The distance to Alpha Centauri can be calculated by the tangent formula: *tangent equals opposite over adjacent.*

$$\text{tangent of parallax} = \frac{\text{distance from earth to sun}}{\text{distance to star}}$$

The earth-sun distance is 1 AU, which provides the baseline of a right triangle. Thus

$$\text{tangent of } 0.75'' = \frac{1 \text{ AU}}{\text{distance to star}}$$

or

$$\text{distance to star} = \frac{1 \text{ AU}}{\tan 0.75''} \approx 360,000 \text{ AU}$$

3. Even the astronomical unit will result in unwieldy numbers when it comes to really distant stars. For this reason, astronomers have defined a larger unit, the *parsec*, as: *the distance of an object whose parallax shift is 1 second of arc.* ("Parsec" is short for *par*allax of 1 *sec*ond.) No star is quite that close, but the parsec is a convenient unit of measure for two reasons:

  i. A star's distance in parsecs is the reciprocal of its parallax in seconds of arc. A star with a parallax of 0.5″ would be 2 parsecs distant. Alpha Centauri, with a parallax of 0.75″, is 1.33 parsecs away. (To review reciprocals, see page 34.)
  ii. One parsec is equal to 3.26 light-years.

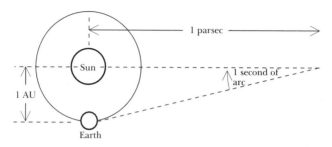

• The Andromeda Galaxy is about 700,000 parsecs away. To convert this to light-years, multiply by 3.26. The result is 2,282,000 light-years, which should properly be rounded to 2 million (see page 23 for a summary of how to round when multiplying measurements). This means that light from the Andromeda Galaxy takes over 2 million years to reach us.

4. For very distant stars, parallax methods give unreliable estimates due to the relatively small size of the baseline and the minuscule angle of parallax. Distances obtained in this way are subject to considerable error, which explains the wide range of star distances given in various star catalogs and textbooks. Because the brightness of a star depends partly on its distance, the same problem comes up in estimates of absolute magnitude. By definition, the absolute magnitude of a star equals the apparent magnitude it would have when viewed from a distance of 10 parsecs. The nearest star cluster, Alpha Centauri, although its apparent magnitude is close to 0, has an absolute magnitude of about 4.4. (The lower the magnitude, the brighter the star; a unit decrease in magnitude translates into an increase in brightness by a factor of 2.5. See page 50 for more details.) The sun, with an apparent magnitude of −26.7, is in fact no brighter than Alpha Centauri; if viewed at a distance of 10 parsecs, it would appear to have a magnitude of 4.7.

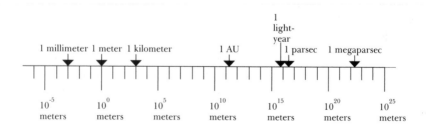

Table 4.9. Summary of Astronomical Distance

1 AU = mean distance from the earth to the sun
  = 149,597,870 km
  ≈ 92,555,800 miles

1 parsec = distance of an object whose parallax shift is 1 second of arc
  = 3.26 light-years
  ≈ 3.086×10^{13} km
  ≈ 1.917×10^{13} miles
  ≈ 206,265 AU

1 kiloparsec = 1000 parsecs

1 megaparsec = 1,000,000 parsecs

1 light-year = distance light travels in one year through a vacuum
  = 5.8785×10^{12} miles
  = 9.46053×10^{12} km
  = 63,240 AU

Sound

Acoustics, the science of sound, is a complex subject that has contributed various words to the language. Like many terms of scientific origin, they are often used imprecisely. *Decibel* is one such term. Most people know that sound is measured in decibels, a scale of numbers that designates levels of sound ranging from the barely audible to the excruciating. Less well known is the fact that decibels are an example of a logarithmic scale based on ratios of sound intensity. This is explained below.

Though the physics of sound is rich in mathematical material, this section concentrates on just a few interesting applications of logarithms and ratios. (Before reading it, you may wish to review logarithms, ratios, and proportions in Chapters 1 and 2.)

1. A *watt* (as discussed in Chapter 3) is a standard unit of power representing the rate at which energy is transferred from one form to another. Although commonly thought of as a measure of electrical power, it may be used for any type of energy, including sound.

2. The power of a sound in watts combines with the size of a sound detector to define its intensity. The *intensity* of a sound is its power per unit area (in watts per square meter, for example).

$$I = \frac{P}{S}$$

3. The loudest sounds we are exposed to rarely exceed one watt per square meter. The softest sounds are about a trillionth of a watt per square meter. Because this is such a large difference in magnitude, sound intensity is a good candidate for measurement on a logarithmic scale. On such a scale, the range from one to a trillion, stepped off in powers of 10, is given by the numbers 1 through 12, which are the powers themselves. (Recall that a logarithm of any power of 10 is the power itself. The logarithm of 1 is 0, because $10^0 = 1$; the logarithm of a trillion is 12, because $10^{12} =$ one trillion.)

4. In the 1920s researchers at Bell Telephone labs were trying to measure the amount of decay in signals transmitted over long distances of telephone wire. They compared the power of the input signal to the power of the output signal by taking the ratio of the two, and because these ratios could be quite large, they then took the logarithm of the ratio. As a result they created a new unit of measure which they named the *bel,* in honor of Alexander Graham Bell. The bel is still the standard unit of measure for *comparative* sound level; it measures the ratio between two sound intensities and not their actual sizes. Thus bels provide an index of relative intensity.

5. A difference of 1 bel means: $\log\left(\dfrac{I_1}{I_2}\right) = 1$. That is, if the intensity of one sound (I_1) is ten times greater than the intensity of another (I_2), the difference in intensities works out to one bel.

- If a signal degrades to $1/10,000$ of its original intensity, it has undergone a loss of 4 bels (because $10^4 = 10,000$).

- A signal that is amplified a million times increases by 6 bels (because $10^6 = 1,000,000$).

6. Since the bel represents a wide range of sound intensity, it was divided into *decibels* (abbreviated *dB*). A decibel is $1/10$ of a bel. A difference of 1 decibel means: $10 \times \log\left(\dfrac{I_1}{I_2}\right) = 1$.

An intensity ratio of 100 (meaning that one sound is 100 times as powerful as another) is equivalent to 2 bels (because $100 = 10^2$), or 20 decibels. An intensity ratio of 1000 is equivalent to 3 bels (because $1000 = 10^3$), or 30 decibels. Thus any *difference* between decibel levels corresponds to a *multiplication* of the intensity by a power of 10. If the decibel level increases by 10, the sound intensity is multiplied by 10. If the decibel level increases by 1, the sound intensity is multiplied by

$10^{0.1}$, or approximately 1.259. (This can be found by using the key marked x^y on a calculator; see page 59 for more details.) Table 4.10 illustrates this relationship.

Table 4.10. Decibels and Sound Intensity Ratios

Difference in level	Intensity ratio (I_1/I_2)
0 dB	1.0 (equal intensities)
1 dB	1.259
2 dB	1.585
3 dB	1.995
4 dB	2.512
5 dB	3.162
6 dB	3.980
7 dB	5.012
8 dB	6.309
9 dB	7.943
10 dB	10
20 dB	100
30 dB	1,000
40 dB	10,000

For each decibel increase, the intensity is multiplied by a factor of $10^{0.1}$, or approximately 1.259.

7. Although bels and decibels represent *ratios* of intensities rather than actual intensities, a decibel has come to be used also as a measure of absolute sound intensity. This was done by establishing a standard of sound intensity to which other sound intensities could be compared. To avoid negative numbers on the decibel scale, this standard was set just below the lowest intensity detectable by the human ear—one-trillionth of a watt per square meter. If this intensity is designated I_0, the measure of a sound whose intensity is I_1 is given by

$$\text{decibel intensity of } I_1 = 10 \times \log\left(\frac{I_1}{I_0}\right)$$

Through this formula the decibel became an absolute measure of sound intensity.

8. As noted above, one watt per square meter is the intensity of the loudest sounds to which we are typically exposed. This is a trillion times greater than the intensity of a barely audible sound. Since a trillion is 10^{12}, the ratio of loudest to softest sounds corresponds to a range of 12 bels (because 12 is the logarithm of 10^{12}), or 120 decibels. As Table 4.11

shows, this range spans the sound of a pin drop to the sound of a jet taking off.

Table 4.11. Examples of Sound Intensity Levels

Sound source	Decibels	Intensity in watts per square meter
Imperceptible sound	0 dB	1×10^{-12}
Pin drop	10 dB	1×10^{-11}
Whisper at 1 meter	20 dB	1×10^{-10}
Empty auditorium	30 dB	1×10^{-9}
Typical bedroom	40 dB	1×10^{-8}
Library (recommended)	50 dB	1×10^{-7}
Quiet conversation	60 dB	1×10^{-6}
Freeway traffic	70 dB	1×10^{-5}
Annoyance level	80 dB	1×10^{-4}
Subway train arrival	90 dB	1×10^{-3}
Loud shout	100 dB	1×10^{-2}
Nightclub	110 dB	1×10^{-1}
Jet taking off at 100 meters	120 dB	1
Limit of amplified speech	130 dB	1×10^{1}
Pain threshold	140 dB	1×10^{2}
Severely damaging	150 dB	1×10^{3}

9. Thus far, no mention has been made of *loudness*, the quality that decibels are assumed to measure. This is because decibels measure only the power of a sound, or its intensity (which is the power per unit area), and this is not the same as loudness.

Loudness, unlike intensity, is a matter of judgment. We react to sounds of different intensities as being loud or soft, although we tend to judge some sounds as being louder than others even when their measured intensities are the same. One factor determining this is the type of sound, and particularly its frequency—how high or low it is pitched. A unit of measure for loudness that takes into account the frequency of a sound is the *phon*.

10. Sound travels by creating oscillations in the particles of the conducting medium. These oscillations are measured in units called *hertz*, which describe cycles (or back-and-forth swings) per second. This is what is meant by *frequency*—the speed of the oscillations is perceived by the ear as pitch, high or low. The range of frequencies detectable by the human ear is 20 to 20,000 hertz. At 1000 hertz, the phon scale is numerically identical to the decibel scale. But this relationship changes as the frequency changes. If the intensity of a sound remains constant as its frequency increases (think of a pure tone emanating from a radio and gradually becoming higher in pitch), it generally begins to sound louder, since we tend to perceive higher-pitched sounds as being louder than those of lower pitch at the same intensity. This is what phons are designed to reflect.

The figure below shows this relationship. The curves represent constant levels of loudness as measured in phons. At 1000 hertz (about the pitch of the C two octaves above middle C on the piano keyboard), phon levels are the same as decibels—for example, 80 phons of loudness coincides with 80 dB of intensity. But at a lower pitch of 20 hertz (a few notes lower than the lowest key on the piano), 80 dB registers as only 20 phons. How much less is not made clear by the change in phons, because phons are not a proportional scale.

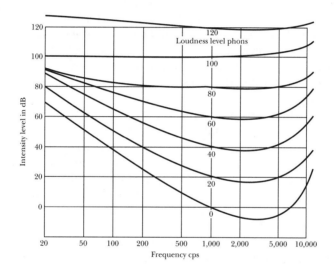

11. Phons are difficult to work with because numerically they do not reflect our subjective perception of loudness. That is, 80 phons of loudness do not sound twice as loud as 40 phons. For this reason we have the *sone,* which is a proportional measure of *perceived* loudness. A sound registering 2 sones sounds twice as loud as one of 1 sone. The relationship between sones and phons is summed up in this formula:

$$S = 2^{(P-40)/10}$$

What this says is that for every 10-unit increase in phons (starting at 40 phons), the perceived loudness doubles. Table 4.12 demonstrates this by giving some sone-phon equivalents.

Table 4.12. Phons and Sones

P	S
40	1
50	2
60	4
70	8
80	16
90	32
100	64
110	128
120	256

12. Noise As has been mentioned, our perception of sound is partly subjective. What we consider loud cannot be described simply in terms of intensity and frequency. We react not merely to the physical attributes of sound but to its pleasantness or unpleasantness. The sound of a low-flying airplane is perceived as louder than the sound of waves crashing on the shore even when the two sounds have about the same frequency and intensity.

Noise, then, is the product of loudness and annoyance. For the purpose of regulating environmental noise, both in the workplace and in our neighborhoods, governmental agencies have steered clear of sones and phons and stayed with the more familiar decibel, by using an adjusted scale. This "A scale" uses the symbol dBA, which can be thought of as decibels adjusted for frequency (and therefore similar to phons). Legal limits are based on units measured in dBA's, and take into account the type of noise (with internal-combustion engines especially targeted) and the time of day. One example of the regulation of workplace noise exposure, created by the Occupational Safety and Health Administration, is given in Table 4.13.

Table 4.13. Permissible Daily Exposure (OSHA)

Sound level (dBA)	Maximum duration
120	0 min.
115	15 min.
110	30 min.
105	1 hr.
102	1.5 hr.
100	2 hr.
97	3 hr.
95	4 hr.
92	6 hr.
90	8 hr.

13. Like light, sound diminishes rapidly with distance traveled. This follows from the *inverse-square law* (see page 90). In an open space, where sounds cannot bounce back, their intensity is inversely proportional to the square of the distance from the source. For example, a gunshot at 15 yards is ⅑ as loud as the same noise heard from 5 yards. This is because it is three times the distance, and the reciprocal of the square of 3 is ⅑. The reason that environmental noise poses such a problem, despite this rapid diminishment, is that we encounter very few open spaces. In most environments, sounds bounce off many surfaces, and this reverberation sustains the noise and offsets the inverse-square law.

5 Probability, Gambling, and Statistics

While mathematics is generally thought to be the most exacting of the sciences, one branch of it is entirely consumed by doubt. This is probability—the study of uncertainty. It occupies a place between the theoretical and real worlds in which almost anything is possible and some things are more probable than others.

Statistics is the branch of mathematics that deals with collecting, analyzing, interpreting, and presenting numerical data. Part of its purpose is to summarize what *has* happened in order to predict what *will* happen or what *might* happen. In this way statistics is closely connected with probability.

Probability and statistics constitute the area of mathematics that is most commonly used (and misused) in everyday life. Together they determine how insurance companies set their rates, how brokers make investment decisions, how casinos reap their profits, and how laboratories test experimental drugs. When a baseball manager sends in a left-handed pinch hitter or an insurance company offers lower rates for nonsmokers, they are playing the laws of averages—using statistics to give themselves the best chance of coming out ahead.

Both probability and statistics employ some technical language, and it is crucial to examine such language closely. When properly used, statistics can clearly summarize complex information; when improperly used, it can bolster a weak argument or create a false impression. As Mark Twain noted in his diary, "There are three kinds of lies: lies, damned lies, and statistics." Thus it is all the more important for the consumer to understand its concepts.

This chapter begins by defining the principles of probability and showing how they apply to games of chance—casino gambling, horse racing, lotteries, and the like. It concludes with an overview of the most frequently encountered statistical concepts. The mathematics is somewhat technical in places, and the reader might want to review certain topics in Chapters 1 and 2, particularly ratios and proportions, percents and decimals, factorials, and calculator calculation. A scientific calculator will be necessary to carry out most of the calculations in the sections that involve permutations and computations.

Probability

1. In the field of probability, an *experiment* is a controlled study whose outcome is uncertain but not entirely unknown. An experiment requires an observer who can note what happens. Every time the observer records the result of an experiment, it is referred to as a *trial*. We all conduct probabilistic experiments: going to the race track or playing the lottery is an experiment, as is purchasing stock, buying insurance, or choosing the best route to commute to work. An individual horse race, lottery drawing, stock performance, or drive home can be thought of as a trial of a probabilistic experiment.

2. An *outcome* is one of the possible results from an experiment trial. It is the basic unit of possible occurrences. In a coin toss, the outcomes are *heads* and *tails*. In the toss of a die (*die* is the singular of *dice*), the outcomes consist of the six sides showing the numbers 1 through 6. In a horse race, the outcomes are the possible orders in which the horses finish.

3. An *event* is some combination of possible outcomes in one experiment trial. If one of the outcomes constituting the event occurs, then the event is said to have occurred.

* In picking a card from an ordinary deck of cards, the outcomes are the 52 cards themselves. Examples of events are:

   "Pick a 4" (one of the four cards numbered 4)
   "Pick a diamond" (one of the 13 cards in the suit of diamonds)
   "Pick a face card" (one of the 12 cards showing a king, queen, or jack)

4. Someone who tosses a die 100 times is conducting 100 trials of the same experiment. The number of times a 1 comes up is called the *fre-*

quency of this outcome. The *relative frequency* of an outcome is the ratio of the frequency of the outcome to the number of trials.

$$\text{relative frequency} = \frac{\text{no. of occurrences}}{\text{no. of trials}}$$

- 100 coin tosses result in 47 heads and 53 tails. The relative frequency of heads is $^{47}/_{100}$, or 0.47. The relative frequency of tails is $^{53}/_{100}$, or 0.53.

5. If the possible outcomes or events in an experiment can be listed, the probability of an outcome or event is its *long-run relative frequency*. If a coin is tossed many times, for example, experiments prove that heads comes up half the time and tails the other half, so the long-run relative frequency of heads is 0.5.

Φ ———————————————————————— Φ

WHAT DOES PROBABILITY MEAN?

All attempts to define probability turn out to go around in circles; probability can only be defined in terms of itself. "Probability" is equated with "likelihood," which is likened to "certainty," which depends on the chance of something happening, which is basically its probability. One way of getting at probability is through the idea of *equally likely events,* such as heads and tails. And yet even if a million coin tosses confirms that heads and tails are equally likely (which, by the way, is very unlikely, as will be explained below), it cannot guarantee the same outcome on the next million tosses. Thus to base the definition of probability on events that are "equally likely" is also circular. Like the geometric terms *point, line,* and *ray,* probability must remain undefined. But this does not prevent us from profiting by it.

Φ ———————————————————————— Φ

TYPES OF PROBABILITY

Probability is a numerical assessment of likelihood. It is expressed as a number between 0 and 1, where 0 means an event is impossible and 1 means it is absolutely certain to happen. Between 0 and 1 lies a range of possible values that rate likelihoods in decimal, fraction, or percent form. A probability of ½ (or 0.5, or 50%) means that an event is expected to occur half the time whenever the experiment is performed. Thus the probability of picking a red card out of a standard deck is ½;

the probability of throwing a total of 13 in a two-dice toss is 0; the probability of tossing a head with a two-headed coin is 1.

Not all statements of probability can be read in the same way. To say that there is a 50% chance that a coin toss will result in a head is not the same as saying that there is a 50% chance of rain tomorrow. In one instance, the experiment is repeatable, and the event (flipping a head) either will or will not happen. Experience shows that it can be expected to happen about half the time. By contrast, a 50% chance of rain means that, based on past experience with similar conditions, there is a 50% chance that *some* rain will fall *somewhere* in the local area at *some* time during the day.

Thus some probabilities are more reliable than others. Predicting the weather is less precise than stating the chances of winning at a roulette table. In the same way, predicting the behavior of the stock market, the outcome of a sports contest, or the chances that a new movie will be a hit are highly speculative. Still, there are some mathematical principles at work in all predictions of likelihood, and these can be summed up in three basic approaches to finding probabilities.

i. *A logical assessment of the physical properties of the experiment* This applies to experiments such as picking a card, guessing a number, flipping a coin, or spinning a dial. It assumes that the basic outcomes are known, are all equally likely, or at least are known to be consistent from trial to trial. In this case, probability is defined as a "proportion of equally probable cases." It assumes that, for example, if six outcomes are possible and each is equally likely, then the probability of any one of them is ⅙. It also assumes that the outcome of any one trial has no effect on the outcome of any other. This is known as the *classical definition of probability.*

ii. *Experiment and observation* When the mechanism at the heart of an experiment is unknown, a large number of trials can reveal its long-run patterns. To prove that a pair of dice is loaded, you could throw them hundreds or thousands or hundreds of thousands of times and observe the frequencies of the results. To determine the chance that a mass-produced object is defective, you could analyze a large number of production runs and count the number of defective units.

iii. *Informed judgment* There are at least two types of situations in which the probability of an event depends on who you ask—that is, in which probability can be a matter of subjective judgment. For example, there are situations that have no precedent, where the most probable outcome can only be found by exercising the best possible judgment based on experience, the available information, and the properties of the experiment. Such is the case with weather forecasting or handicapping horse races, situations in which there is plenty of data but no exact precedents. What has already happened merely *suggests* what to expect next. Another type of subjective probability depends on who knows what. If two people bet on a coin toss, and one of them believes the coin is fair while the other knows it is two-headed, the probability of tossing a head is ½ in the first instance and 1 in the second, respectively.

CALCULATING PROBABILITY

Probability is calculated as a ratio of the probable to the possible. That is, assigning probability is based on the fact that what is *probable* depends on what is *possible,* and exactly *how* probable depends on how many possibilities there are. This leads to the following rules:

1. The First Principle of Probability If an experiment has a set of distinct outcomes, each of which is equally likely, then the probability of an event is the ratio of the number of outcomes to the total number of possible outcomes.

- Picking a card from a deck of cards has 52 distinct outcomes. The event "pick a diamond" consists of 13 of these outcomes, the 13 cards in the suit of diamonds. Thus the probability of picking a diamond is given by

$$P(\text{diamond}) = \frac{\text{no. of diamonds}}{\text{no. of cards}} = \frac{13}{52} = \frac{1}{4}, \text{ or } 0.25$$

2. Probability as Relative Frequency The *relative frequency* of an event is the ratio of the number of occurrences of the event to the number of trials.

$$\text{relative frequency} = \frac{\text{no. of occurrences}}{\text{no. of trials}}$$

In many situations, a large number of trials can reveal what the probabilities of the possible outcomes should be. This is the principle used in computing batting averages, product reliability (and failure rates), and manufacturing defects.

- A check of a production run of 5000 silicon wafers reveals that 143 are defective. Therefore, the probability that a wafer chosen at random is defective would be $143/5000$, or 0.0286.

Such a probability can change in the face of new data. The statistician must decide how reliably the relative frequency from a sample reflects the entire population.

Φ ———————————————————————— Φ

THE LAW OF AVERAGES

One of the most important theorems in probability theory is the *Law of Large Numbers,* which is known informally as the *law of averages.* This is the law that tells you there is a fifty-fifty chance of winning a coin toss. It does *not* say that in 100 coin tosses you should always expect 50 heads and 50 tails. In fact, it is more likely that you will *not* throw the same number of heads as tails. The reason why will be made obvious later in this chapter (see page 204). As for the Law of Large Numbers, it does *not* say that you will, in the long run, throw the same number of heads as tails, but rather that the *ratio* of heads to tails will *approach* 1 to 1. In one million tosses, it would not be surprising to throw 500 more heads than tails. 500,250 heads out of 1,000,000 tosses is a ratio of 0.50025, which is very close to the expected value of 0.5.

This contradicts the widely held notion that events have to even themselves out. It is a common fallacy that if someone tosses five heads in a row, the probability of a tail on the sixth toss is better than fifty-fifty because tails is somehow "overdue," that the odds have to balance out because too many heads have come up. What this overlooks is the Big Picture, which is governed by the law of averages. Runs of two, three, four, five, six, or more heads *and* tails are expected in any long string of tosses. Only the players keep any memory of what has taken place; the coin does not. If you throw 60 heads and 40 tails in 100 tosses, the chances are very high that after 200, 300, or 400 tosses, heads will still be running ahead of tails. What you *can* expect is that as more and more tosses pile up, the ratio of heads to tails will get closer to 1 to 1.

Φ ———————————————————————— Φ

3. The Probability That an Event Will Not Happen In many instances, finding the probability that something will *not* happen is easier than finding the probability that it will. These two probabilities must add up to one. Thus finding the probability of an event is the same as subtracting the probability that it *won't* happen from 1.

Probability that A will occur = 1 − (probability that A will not occur)

• The probability of getting *at least* one tail in a two-coin toss can be found by calculating the probability of getting *no* tails, which is the only other possibility. Getting no tails means tossing head-head, which has a probability of ¼ (see below). Thus the probability of at least one tail is 1 − ¼ = ¾.

INDEPENDENT AND MUTUALLY EXCLUSIVE EVENTS

1. When an experiment is made up of several trials, as in the case of a multiple coin toss, the trials combine to produce a single result. We say that each trial is *independent* of the others if the outcome of one trial has no effect on the outcome of any other. When a coin is tossed, the result has nothing to do with what happened on the previous toss and will not affect the next one. If someone managed to throw nine heads in a row, the chance of a head on the next toss is still an even bet. As mathematicians say, "The coin has no memory."

2. To calculate the probability that two or more independent events will occur, multiply their probabilities.

• If two coins are tossed simultaneously, what is the probability that they will both come up heads? (The problem remains the same if one coin is tossed twice in succession.) Because the events are independent, the probability of two heads, simultaneously or in a row, is found by multiplying the probability of one head times itself: ½ × ½ = ¼.

• The probability of three heads in a row is $\frac{1}{2} \times \frac{1}{2} \times \frac{1}{2} = \frac{1}{8}$.

• The probability of a given number (designated n) of heads in a row is $(\frac{1}{2})^n$ or $\frac{1}{2^n}$.

• The probability of picking two aces out of an ordinary deck must account for the fact that the first ace picked is not replaced. The probability of picking one ace is $\frac{4}{52}$. This leaves three aces in a 51-card deck. The probability of picking the second ace is $\frac{3}{51}$. Thus the probability of picking two aces at random out of a well-shuffled deck is $\frac{4}{52} \times \frac{3}{51} = \frac{12}{2652} = \frac{1}{221} \approx 0.0045$.

3. Two or more events are said to be *mutually exclusive* if no two of them can possibly happen in the same trial. When picking one card from a deck, it is impossible to pick a card that is both a spade and a club; those events are mutually exclusive. If two or more events are mutually exclusive, the probability that either one or the other will occur is the sum of their probabilities.

• The probability of drawing either an ace or an even-numbered card from a standard deck is calculated by adding their respective probabilities:

$$P(\text{ace}) = \frac{4}{52}$$

$$P(\text{even-numbered card}) = \frac{20}{52}$$

$$P(\text{ace } or \text{ even-numbered card}) = \frac{4}{52} + \frac{20}{52} = \frac{24}{52}$$

where P stands for "probability of."

Once again: If events A and B are independent, the probability that A *and* B occur is the *product* of their individual probabilities. If events A and B are mutually exclusive, the probability that A *or* B occurs is the *sum* of their probabilities.

Counting Methods

Probability applies to situations in which all the possible outcomes of an experiment are known. This explains why the study of probability began with the study of gambling. In all games of chance, the possibilities are completely known: a deck has 52 cards, a die has 6 sides, a roulette wheel has 38 numbers.

In such situations, the probability of an event can be found by counting the number of ways something can happen as well as the total number of things that can happen. In most games of chance this involves very large numbers, so many people play hunches instead of taking advantage of calculable odds. The sheer number of possible outcomes makes it difficult to calculate probabilities mentally, and the game often unfolds very fast. Some familiarity with the principles of counting could help in these situations.

For example, imagine a lottery game that involves matching five numbers. To find the probability of winning, it is necessary to calculate how many five-number arrangements are possible, which involves counting. But because so many arrangements are possible, some counting method is needed. The method will involve nothing more than multiplication and division, but the way it is done will depend on how the arrangements are made. If the order of the five numbers is important, the arrangement is called a *permutation*. If order is unimportant, it is a *combination*. If any number can be used more than once, and order is important, the possible orderings are referred to as *assortments*. All these methods of counting rely on a basic rule called the principle of multiplication.

THE PRINCIPLE OF MULTIPLICATION

If a group of objects is to be assembled by choosing one object at a time, the number of ways of selecting the group is found by multiplying the number of possible choices for each position.

Total no. of ways of choosing = (no. of choices for 1st position) × (no. of choices for 2nd position) × (no. of choices for 3rd position) × . . . × (no. of choices for last position)

This is called the *principle of multiplication*.

Any event made up of two or more outcomes is called a *compound event*. In compound events, the principle of multiplication may be used to determine the number of possible outcomes. Tossing two dice, for example, is a compound event. There are 6 ways for the first die to come up and 6 ways for the second. The number of possible outcomes is $6 \times 6 = 36$. Tossing a die and then flipping a coin is also a compound event. Because there are 6 ways for the die to come up, and 2 ways for the coin, the number of possible die-coin outcomes is $6 \times 2 = 12$.

PERMUTATIONS

1. A *permutation* is an ordering of objects in which each object appears only once and *order is important*. The six possible permutations of the letters A, B, and C, for example, are ABC, ACB, BAC, BCA, CAB, CBA.

2. The number of permutations of three objects can be found by noting that there are 3 choices for the first position, 2 for the second, and 1 for the third. By the principle of multiplication, the number of possible permutations of three objects is $3 \times 2 \times 1$, or 6.

3. The expression $3 \times 2 \times 1$ is called "three factorial." This is normally written with an exclamation point standing in for the word *factorial*. Thus $3! = 3 \times 2 \times 1 = 6$.

4. In general, the number of permutations of n objects is $n!$ Ten people can be lined up in 10! ways. This amounts to $10 \times 9 \times 8 \times 7 \times 6 \times 5 \times 4 \times 3 \times 2 \times 1$, or 3,628,800 ways. In a horse race involving seven horses, there are 7!, or 5,040, possible orders of finish. Table 5.1 shows how quickly the factorials of the counting numbers grow. (For more on factorials, see page 51.)

Φ ——————————————————————————— Φ

THE BIRTHDAY PROBLEM

Here is one of the most popular applications of the laws of probability. There are many ways to set it up, but in any form it is always known as the birthday problem.

Would you be willing to make an even-money bet that two children in a classroom with 26 children share the same birthday?

Setting aside the moral issue of wagering in front of children, we treat this as a scientific investigation and begin by asking each child his or her birthday. The first child's birthday might fall on any day of the year (we will ignore leap years and use a 365-day year). The probability that the second child's birthday is different is $^{364}/_{365}$. Notice that we concentrate on the probability that there is *no* match; this makes the problem easier.) After the first child, there are 364 birthdays left that result in no match. After the second child, there are 363 days left, so the probability that the third child's birthday does not match either of the first two is $^{363}/_{365}$.

Each of these events is independent of the others. Consequently, to find the probability that there are no matching birthdays among these three children, we multiply the probabilities:

$$\frac{364}{365} \times \frac{363}{365} \approx 0.9918$$

So far, the chance of no matches is almost certain. But by the tenth child the probability of no matches is:

$$\frac{364}{365} \times \frac{363}{365} \times \frac{362}{365} \times \frac{361}{365} \times \frac{360}{365} \times \frac{359}{365} \times \frac{358}{365}$$
$$\times \frac{357}{365} \times \frac{356}{365} \approx 0.8831$$

If we continue to the twentieth child, the probability of no match drops to 0.5886. The surprising result is that the fifty-fifty cutoff is at 23. If the group has 23 people, it is worth betting even money on two birthdays coinciding because it has better than a 50% chance of being true. By the 26th child the probability of no match is down to 0.4018, which leaves close to a 60% chance of matching birthdays. In a classroom with 30 students, your odds of a match are better than 70%.

Φ ——————————————————————————— Φ

Table 5.1. Factorials

n	$n!$
0	1
1	1
2	2
3	6
4	24
5	120
6	720
7	5,040
8	40,320
9	362,880
10	3,628,800
11	39,916,800
12	479,001,600
13	6,227,020,800
14	87,178,291,200
15	1,307,674,368,000

5. The number of permutations of the three letters A, B, and C is 6, as we have seen. But how many three-letter permutations are possible using all the letters of the alphabet? The number of permutations of three objects chosen from a group of 26 objects can be found using factorials. The answer in this case is $26!/23!$, or 15,600. In general, the number of permutations of r objects chosen from a set of n objects is designated $P_{n,r}$ or $_nP_r$, and is given by the formula:

$$P_{n,r} = \frac{n!}{(n-r)!}$$

In words, if you have n objects, and you want to know how many permutations you can make by choosing r of them, find the difference between n and r, find the factorial of this number, and divide it into n factorial.

• A disk jockey has to play 9 songs from a list of 15 songs during an hour-long broadcast. The number of possible song combinations (in which order is important) is given by $P_{15,9}$.

$$P_{15,9} = \frac{15!}{(15-9)!} = \frac{15!}{6!}$$

$$= \frac{15 \times 14 \times 13 \times 12 \times 11 \times 10 \times 9 \times 8 \times 7 \times 6 \times 5 \times 4 \times 3 \times 2 \times 1}{6 \times 5 \times 4 \times 3 \times 2 \times 1}$$

$$= 1,816,214,400 \text{ possible playlists.}$$

6. Notice that zero factorial is defined to equal one ($0! = 1$). This means that the number of permutations of all n objects (written $P_{n,n}$) would be $n!/0!$, which equals $n!$ On the other hand, the formula tells you that there is only one way to choose zero objects out of n objects: $P_{n,0} = n!/n! = 1$.

♦ In a 12-horse field, the outcomes consist of all possible orders of finish. This would be $P_{12,12} = 12! = 479,001,600$. What matters to bettors are the win-place-show possibilities, which consist of the number of permutations of 3 objects chosen from 12. This is given by $P_{12,3}$, which is $12!/9!$, or 1320. (Another approach is to note that there are 12 candidates for the winner, 11 for second place, and 10 for third place. By the principle of multiplication, the number of possibilities turns out to be $12 \times 11 \times 10$, or 1320.)

COMBINATIONS

1. In a *combination*, unlike a permutation, *order does not matter.* There is only one combination of the letters A, B, and C, for example: ABC, CBA, and BAC all represent the same combination of letters. The only interesting combinations are those that result when objects are chosen out of a group of objects. As with a permutation, no object in a combination can be used more than once. A poker hand, for example, represents a combination of 5 objects chosen from a set of 52 objects. If the order of the 5 cards is rearranged, it is still the same hand and thus the same combination.

Because order is unimportant in a combination, there are fewer combinations than permutations of a given number of objects. If 5 basketball players are chosen from an 8-man roster, the number of *permutations* would be

$$\frac{8!}{(8-5)!} = \frac{8!}{3!} = 8 \times 7 \times 6 \times 5 \times 4 = 6720.$$

Because the ordering of the 5 players doesn't matter, some of these permutations are simply rearrangements of the same 5 players. The number of these duplications is $5!$, or $5 \times 4 \times 3 \times 2 \times 1$, which is the number of ways of arranging 5 players. So the number of *combinations* of 5 players on an 8-man roster is found by dividing the number of permutations of the players by the number of duplications of 5 objects. This would be

$$\frac{8 \times 7 \times 6 \times 5 \times 4}{5 \times 4 \times 3 \times 2 \times 1} = 56$$

2. In general, the number of combinations (C) of r objects taken from a set of n objects, in which an object can be used just once (and order is unimportant), is designated $C_{n,r}$ or $_nC_r$, where

$$C_{n,r} = \frac{P_{n,r}}{r!} = \frac{n!}{r! \times (n - r)!}$$

That is, to find the number of groups of size r that can be chosen from n objects, where order doesn't matter, find r factorial, multiply it by the factorial of $(n - r)$, and divide this product into n factorial. Or, find $P_{n,r}$ and divide by $r!$

3. Another notation for $C_{n,r}$ is $\binom{n}{r}$, which reads "n choose r," and which means, "the number of ways of choosing r objects out of a set of n objects."

♦ The number of ways that a committee of 3 can be chosen from a group of 15 people is

$$\binom{15}{3} = \frac{15!}{3! \times (15 - 3)!} = \frac{15!}{3! \times 12!} = \frac{15 \times 14 \times 13}{3 \times 2 \times 1} = 455$$

♦ The number of foursomes that can be put together from a gathering of 14 golfers is

$$\binom{14}{4} = \frac{14!}{4! \times 10!} = \frac{14 \cdot 13 \cdot 12 \cdot 11}{4 \cdot 3 \cdot 2 \cdot 1} = 1001$$

ASSORTMENTS

A group of objects is selected from a larger group in such a way that an object can be used *more than once*. In this case, the number of possible groupings is neither a permutation nor a combination. Writing a number, for example, involves writing a string of digits. Any digit can be used any number of times. Choosing a password involves writing a string of letters of the alphabet, with repetition of letters allowed. Because the digits and letters can be used more than once, this situation requires a special counting method.

The number of arrangements, or *assortments*, of numbers that can be made from 10 digits, or of passwords that can be made from the 26 letters of the alphabet, is found using the principle of multiplication.

no. of assortments = (no. of choices for 1st position) × (no. of choices for 2nd position) × . . . × (no. of choices for last position)

♦ To find how many three-digit numbers there are, begin with the 10 possible choices for the first digit, multiply by the 10 possible choices for the second digit, then by the 10 possible choices for the third digit. Thus there are $10 \times 10 \times 10$, or 1000 such arrangements. These include numbers that begin with 0, such as 001, 015, and 090. The entire list includes all of the counting numbers from 000 to 999.

♦ The number of possible license plates containing three letters followed by three digits is: $26 \times 26 \times 26 \times 10 \times 10 \times 10 = 17,576,000$. This is because there are 26 choices for each of the 3 letters and 10 choices for each of the 3 digits.

Φ ———————————————————— Φ

ASSORTMENTS, PERMUTATIONS, AND COMBINATIONS: A REVIEW

Assortments: If an object can be used more than once, use the principle of multiplication. The number of ways is equal to:

(no. of choices for 1st position) $\times$ (no. of choices for 2nd position) $\times$ . . . $\times$ (no. of choices for last position)

Permutations: If an object cannot be used more than once, and order *is* important, the number of ways is equal to:

$$P_{n,r} = \frac{n!}{(n - r)!}$$

Combinations: If an object cannot be used more than once, and order *is not* important, the number of ways is equal to:

$$C_{n,r} = \frac{n!}{r! \times (n - r)!}$$

Note: Most scientific calculators have not only a key for factorials (usually marked x!), but keys for permutations and combinations as well. These are usually marked *nPr* and *nCr*. For example, to calculate the number of *permutations* of 4 objects chosen from a set of 9 objects, press $\boxed{9}$ $\boxed{nPr}$ $\boxed{4}$. The result is 3024, which is 9! divided by $(9 - 4)!$ To calculate the number of *combinations* of 4 objects chosen from a set of 9 objects (that is, "9 choose 4"), press $\boxed{9}$ $\boxed{nCr}$ $\boxed{4}$. This gives the result 126, which is 9! divided by the product of 4! and $(9 - 4)!$

Φ ———————————————————— Φ

PASCAL'S TRIANGLE

Blaise Pascal, the seventeenth-century French mathematician, turned his passionate interest in gambling into a science and founded the modern theory of probability. Among his many contributions to mathematics, he came up with an intriguing triangle of numbers that is easy to construct, has profound applications, and bears his name.

Here is how to construct Pascal's triangle: Begin with a triangle made of three 1's.

$$1$$
$$1 \quad 1$$

Add as many rows below as desired. Each new row must begin and end with 1, but the numbers in between are found by summing the two numbers above. This is what results:

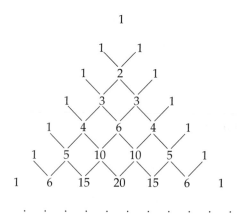

(You may recognize the triangle as the pattern of the *binomial coefficients* from second-year algebra. The third row, for example, consists of the coefficients of $(x + 1)^2$ when it is multiplied out as $x^2 + 2x + 1$. The fourth row consists of the coefficients of $(x + 1)^3$, which multiplies out to $x^3 + 3x^2 + 3x + 1$; the fifth row gives the coefficients of $(x + 1)^4$, or $x^4 + 4x^3 + 6x^2 + 4x + 1$; and so on.)

The triangle also gives the probabilities related to coin tosses, and thus connects to combinations. To show this connection, Table 5.2 writes out Pascal's triangle in a slightly different form.

Written in the form of a table, the numbers of the triangle are easier to read as the values of the combinations $C_{n,r}$, or "n choose r." To find the number of combinations of 3 objects chosen from a set of 6 objects, for example, read down the 3 column, and across in the 6 row to find the

Table 5.2. Combinations and Pascal's Triangle

n/r	0	1	2	3	4	5	6	
0	1							$= 2^0 = 1$
1	1	1						$= 2^1 = 2$
2	1	2	1					$= 2^2 = 4$
3	1	3	3	1				$= 2^3 = 8$
4	1	4	6	4	1			$= 2^4 = 16$
5	1	5	10	10	5	1		$= 2^5 = 32$
6	1	6	15	20	15	6	1	$= 2^6 = 64$
. . .								. . .
n								$= 2^n$

value 20. This is the value of "6 choose 3." The numbers in the last column are the sums of each row.

How is it used? One application involves the probabilities of getting 0, 1, 2, 3, 4, and 5 heads in 5 coin tosses. By the principle of multiplication, there are $2 \times 2 \times 2 \times 2 \times 2 = 32$ possible outcomes in a 5-coin toss. One possibility is 5 heads, which can happen in only one way, and therefore has a probability of $\frac{1}{32}$. Another outcome is all tails (or no heads), which also has a probability of $\frac{1}{32}$. There are exactly 5 ways to get 1 head or 1 tail (that is, on the first, second, third, fourth, or fifth toss); thus the probability of one head or one tail is $\frac{5}{32}$. Now look at the $n = 5$ row in the table. It sums to 32, and the numbers seem to coincide with the pattern of probabilities in the coin experiment. This is not a coincidence; the numbers in the following list make clear the connection between the $n = 5$ row in Pascal's triangle and the frequency of heads (or tails) in a 5-coin toss:

$$P(0 \text{ heads}) = \frac{1}{32}$$

$$P(1 \text{ head}) = \frac{5}{32}$$

$$P(2 \text{ heads}) = \frac{10}{32}$$

$$P(3 \text{ heads}) = \frac{10}{32}$$

$$P(4 \text{ heads}) = \frac{5}{32}$$

$$P(5 \text{ heads}) = \frac{1}{32}$$

Pascal's triangle can tell you how many times you can expect equally likely events to be out of balance. In a repeated coin toss, you might expect the number of heads to equal the number of tails most of the time. You might also assume that most families with four children have two boys and two girls, and that most six-child families have three and three. Yet neither is the case, even though the individual probabilities—head/tail, boy/girl—are even.

If you consider the $n = 6$ row of Pascal's triangle, it says that the probability of getting 3 heads in a 6-coin toss is $20/64$, which is less than a third. This means that two-thirds of the time you will not get the same number of heads as tails (or boys as girls). If you toss a coin 100 times, you could calculate the probability of 50 heads and 50 tails by generating the $n = 100$ row of Pascal's triangle, or, more simply, by calculating $C_{100,50}$ on a calculator (see page 201). If you then divide this by 2^{100}, which is the number of possible outcomes, you will find that the probability of 50 heads and 50 tails in 100 tosses is about 0.08, or 8 times out of 100 trials.

In other words, in a series of repeated trials of an experiment, *the most probable event is not automatically the most likely to happen.* Of all the boy-girl distributions possible in a 10-child family, 5 boys and 5 girls is the most probable *single* outcome. But it is less likely than all of the other possibilities combined. Most ten-child families have an unequal number of boys and girls.

CARD GAMES

A standard deck of cards contains 52 cards divided into four suits: spades, hearts, diamonds, and clubs. Each suit consists of 13 cards labeled Ace, 2, 3, 4, 5, 6, 7, 8, 9, 10, jack, queen, and king. (Not all card games use this deck—some use only a portion of it, others use more than one deck—but only games that involve drawing cards from a single, standard deck will be considered here.)

1. The probability of drawing any particular card from a standard deck is 1 in 52. This assumes that the order of the deck is completely random. Studies have shown that at least five standard shuffles are necessary to randomize a deck; seven shuffles is safer.

2. With every card dealt from a deck, the number of cards that remain is reduced by one. This will change subsequent probabilities. Consider the probability of dealing four consecutive aces. The probability that the first card will be an ace is 4 chances out of 52, or $4/52$. Assuming an ace is dealt, there are now 3 aces out of the 51 cards remaining in the deck, and the probability that the next card is an ace is $3/51$. If this results in an ace, there would then be 2 aces remaining in a 50-card deck; thus the probability of dealing a third ace would be $\frac{2}{50}$. As for the fourth card, the probability of dealing an ace once three aces have already been dealt is $\frac{1}{49}$. Thus the probability of dealing four aces from a well-shuffled deck is $\frac{4}{52} \times \frac{3}{51} \times \frac{2}{50} \times \frac{1}{49} = \frac{1}{270,725} \approx 0.0000037$.

3. Probabilities associated with card games vary considerably in complexity. The simplest are those associated with specific card combinations such as poker hands. This involves selecting or dealing five cards facedown. Once the cards have been viewed, the probabilities of subsequent draws will change.

Table 5.3 shows a list of poker hands, in order from best to worst, with the number of ways they can be dealt and their probabilities of being dealt. A *pair* is two cards of the same denomination (4's, jacks, etc.). *Three of a kind* is three cards of the same denomination. A *straight* is five cards (of any combination of suits) in sequence. A *flush* is any five cards of the same suit. A *straight flush* is five cards of the same suit that are also in sequence. A *full house* is a pair plus three of a kind. A *royal flush* is a straight flush consisting of ace, king, queen, jack, and 10.

Table 5.3. Poker Hand Probabilities—Odds Against Various Pat Hands

Hand	No. of ways	Probability	Expected frequency
Royal flush	4	0.00000154	1 in 649,740 hands
Straight flush	36	0.00001385	1 in 72,193 hands
Four of a kind	624	0.0002401	1 in 4165 hands
Full house	3744	0.0014406	1 in 694 hands
Flush	5108	0.0019654	1 in 509 hands
Straight	10,200	0.0039246	1 in 255 hands
Three of a kind	54,912	0.0211285	1 in 47.33 hands
Two pair	123,552	0.0475390	1 in 21 hands
Pair	1,098,240	0.4225690	1 in 2.37 hands
None of the above	1,302,540	0.5011774	1 in 2 hands
Totals	2,598,960	1.0000000	

This table shows that there are 2,598,960 possible 5-card combinations in a deck of 52 cards. (That is, $_{52}C_5$, or $C_{52,5}$, or "52 choose 5.") The probability of any one hand is the number of ways the hand can be dealt, divided by the number of possible combinations. Of the 40 possible straight flushes, for example, only 4 are royal flushes, so the probability of getting a royal flush is 4 out of 2,598,960, or about 0.00000154. Any hand that can be dealt must fall into one and only one of the categories above. Thus, although a straight flush is both a straight and a flush, it does not count in either of those categories, just as royal flushes are not included among straight flushes.

Φ ———————————————————————— Φ

ORIGINS OF THE MODERN DECK
OF PLAYING CARDS

When Johannes Gutenberg printed a deck of tarot cards in 1440 (the same year he printed his first Bible), playing cards were already well-known in Europe. Thought to have been brought back from the Middle East by returning crusaders, or perhaps from China by Marco Polo, cards became increasingly popular as they became more widely available, and the deck itself evolved within a century to the deck we use today.

The tarot deck that Gutenberg produced contained 78 cards—four denominations of 14 cards each, plus a set of 22 cards known as *atouts* (meaning "assets"), which served as trump cards. The four denominations, or suits, represented the social classes: swords stood for the nobility, coins for merchants, batons (or clubs) for peasants, and cups (or chalices) for the clergy. Each suit contained cards running from 1 to 10, as well as a king, queen, knave, and cavalier.

By the year 1500, the French had dropped all of the atouts except for *le fou* (the joker). They also dispensed with the cavalier, thus leaving 52 cards and a pair of jokers. As for the denominations, their names and symbols have come down to us through a haphazard process. The French names for the suits are *piques* ("pikes") for spades, *carreaux* ("tiles") for diamonds, *trèfles* ("trefoil" or "cloverleaf") for clubs, and *coeurs* ("hearts"). Our *spade* comes from the Spanish for "sword," *espada*. Diamonds evolved from the tarot's coin suit; the French converted the symbol into a diamond shape. Clubs survive from the tarot deck, although the symbol was stylized by the French. Finally, the transition from cup to heart is obscure, but may simply have resulted from a gradual transformation of symbols.

Φ ———————————————————————— Φ

4. In contract bridge, each of the four players receives 13 cards. There are "52 choose 13"—or 635,013,559,600—different hands, all of them equally likely. Thus, the seemingly extraordinary event of being dealt a hand consisting of all spades is no more unusual than being dealt any other hand.

Φ ———————————————————————— Φ

LORD YARBOROUGH'S WAGER

A *Yarborough* is a bridge hand containing no card above a 9. Because aces, kings, queens, jacks, and 10's are referred to as *honors* in bridge, the Yarborough is a hand with no honors. It is so weak a hand that it could be said to have no honor.

Charles Anderson Worsley, the Second Earl of Yarborough, an otherwise undistinguished nineteenth-century lord, thought a bridge hand with no honors to be so amusing that he maintained a standing offer to pay £1000 to anyone who would put up £1 on the prospect of being dealt one. Because there are 32 cards (8 in each of the four suits) that can be part of a Yarborough, the number of possible Yarboroughs is "32 choose 13," or 347,373,600. There are 635,013,559,600 possible bridge hands. Consequently, the probability of being dealt a Yarborough is $347,373,600/635,013,559,600$, which works out to a mere 0.000547.

For the record, it should be noted that Lord Yarborough did not offer a very fair bet. Because the hand is expected to occur once in 1828 hands dealt, His Lordship should have offered £1827 to £1 to make it fair, or at least honorable.

Φ ———————————————————————— Φ

PROBABILITIES WITH DICE

Dice were invented for the express purpose of gambling. In fact dice games have so little intrinsic interest that in the absence of wagering they would hardly be worth playing. But the six-sided die and the games and gambling strategies it has inspired provide such useful illustrations of classical probability theory that dice have a prominent place in all introductory probability textbooks.

The most popular dice game is craps, a game whose richness of terminology has added many colorful words to the English language (as have bridge and poker). Another dice game, once popular but now out of favor, is chuck-a-luck. Both are discussed in detail in the next section. This section provides an introduction to dice probabilities.

1. In tossing a fair die in a fair way, each of the six sides is equally likely. The die has no memory, so each toss carries the same odds: any one of the six sides has a probability of ⅙.

2. A two-dice toss, such as occurs in craps, is simultaneous, although the probabilities would be the same if the two dice were tossed one after the other. To avoid confusion in counting the possible outcomes, imagine that one die is red and the other green. The outcome on the red die is independent of the outcome on the green die, so probabilities can be calculated by multiplication. For example, the probability that a 1 will be thrown on each die is ⅙ time ⅙, which equals ⅟₃₆. A total of 4 can occur in any of three ways—1 and 3, 2 and 2, 3 and 1 (notice that 1 and 3 is different from 3 and 1). Each of these occurrences has a probability of ⅟₃₆, and they are mutually exclusive. Thus the probability of one of them occurring is found by summing their probabilities: ⅟₃₆ + ⅟₃₆ + ⅟₃₆ = ³⁄₃₆, or ⅟₁₂.

3. The process described above can be sidestepped by showing all of the possible two-dice combinations, as in the figure below.

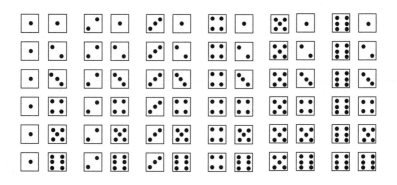

Table 5.4, which is based on this figure, shows how many ways each of the numerical totals can occur. A 7, for example, can be thrown in six ways. Thus its probability is ⁶⁄₃₆, which means that a 7 occurs once in every six throws, on average. Notice that the sum of all the probabilities in the table is 1; this is because each outcome is mutually exclusive, and the table covers all of the possible outcomes.

Table 5.4. Two-Dice Probabilities

Two-dice total	2	3	4	5	6	7	8	9	10	11	12
Probability	⅟₃₆	²⁄₃₆	³⁄₃₆	⁴⁄₃₆	⁵⁄₃₆	⁶⁄₃₆	⁵⁄₃₆	⁴⁄₃₆	³⁄₃₆	²⁄₃₆	⅟₃₆

4. Three-Dice Probabilities With three dice there are more outcomes. In the three-dice game of chuck-a-luck, the dice are marked either with spots or with symbols (usually clubs, diamonds, hearts, spades, crowns, and anchors). Each player bets on one of the symbols. Payoffs are based on how many matches there are in one toss. There are 216 possible outcomes ($6 \times 6 \times 6 = 216$). Three symbols matching the symbol bet on is the least likely outcome, of course, because there is only one way it can happen; thus its probability is $\frac{1}{216}$. The game is explained in detail on page 216.

Odds and Mathematical Expectation

The term *odds* can refer to the probability that an event occurs, or it can be used to indicate the payoff on a winning bet. Since these meanings do not always amount to the same thing, the term is somewhat ambiguous. In its everyday sense the word *odds* is synonymous with *chances*, but these are not necessarily the same as the true probability of an event, as a simple example will show.

A coin toss has a 50% chance of coming up heads and a 50% chance of tails. These are referred to as *1-to-1 odds*, or *even odds*. If a casino were to pay even odds on such a game, it would pay one dollar for each dollar bet. But chances are it wouldn't. It might instead pay 95 cents, which would give it an advantage on each play. In that case it would be offering odds of 95 to 100. Thus the "true odds" are not always the same as the "payoff odds."

DEFINITIONS

1. The probability that an event A will occur is denoted $P(A)$ (that is, "probability of A"). The probability that an event A will not occur can be denoted $P(notA)$. Because event A either occurs or does not occur (either something happens or it doesn't), then $P(A) + P(notA) = 1$. (Recall that 1 means the absolute certainty of something occurring.) Thus the probability that event A does *not* occur is $1 - P(A)$.

2. The *odds* that an event occurs is the ratio of the probability that it *will* occur to the probability that it *will not* occur. This is referred to as the *odds for* the event.

$$\text{odds for event } A = \frac{P(A)}{1 - P(A)} = \frac{\text{probability that } A \text{ occurs}}{\text{probability that } A \text{ does not occur}}$$

• In tossing two dice, the probability of throwing a total of 7 or 11 is $\frac{2}{9}$. Thus the *odds for* throwing 7 or 11 would be $\dfrac{\frac{2}{9}}{1 - \frac{2}{9}}$, or $\frac{2}{7}$. This would be called 2-to-7 odds.

If computing the odds using the formula above seems difficult, it is because it involves a complex fraction (a fraction within a fraction). An easier approach involves calculating odds as the ratio of occurrences to nonoccurrences in a large number of trials.

• In 100 coin tosses, the odds of (or *for*) getting a tail would be the ratio of the expected number of tails thrown to the number of heads thrown. The odds would be 50 to 50, which is the same as 1 to 1, or even odds.

• In 36 two-dice tosses, on average, a total of 7 will occur six times and a total of 11 will occur twice, for a total of eight occurrences of either 7 or 11. Thus a 7 or 11 will occur eight times and *not* occur 28 times in 36 trials, making a ratio of 8 to 28 occurrences vs. non-occurrences, which is the same as 2 to 7. These are the odds for 7 or 11 on any given throw in the game of craps.

3. If the *odds for* an event are given as a to b, then the probability of that event would be $\dfrac{a}{a+b}$.

• If a bookie lays odds of 3 to 5 for a horse to win, he is placing the horse's probability of winning at $\dfrac{3}{3+5}$, or ⅜. In this case, if the horse wins, a \$3 bet should return \$5 in winnings (plus the return of the original \$3). The payoff would be \$8.

4. The *odds against* an event are the reciprocal of the *odds for* the event. Odds of 3 to 5 *for* an event translate into odds of 5 to 3 *against* it. In gambling, the standard practice is to list the odds *against* a wager rather than the odds *for* it.

• A gambler places a \$5 bet on a wager that carries odds of 20 to 1 against. In betting parlance this is called a *20-to-1 shot*. It converts to a probability of ²⁰⁄₂₁ *against* or ¹⁄₂₁ *for* a win. With these odds, a \$1 bet would result in a \$20 profit (plus the return of the \$1 stake). A \$5 bet would reap a \$100 profit.

5. True Odds and House Odds In most betting situations, the person or establishment setting the payoffs on bets always seeks an advantage. This is done by making the payoff less than the actual (or true) odds would dictate. The simplest example is a coin toss. As noted above, the probability of success with a fair coin is ½, which is the same as 1-to-1 odds. Thus a successful $1 bet should earn $1 in winnings. If the house will only pay 95 cents on a $1 bet, they have an advantage of 2.5 cents per bet (since they will win half the time). Thus the *house odds* would be 95 to 100 against even though the *true odds* are 1 to 1. Every casino game carries a house advantage, which means that the house odds are set less than the true odds. This assures that, in the long run, every casino game is a losing proposition for the bettor.

Φ ——————————————————————— Φ

A WARNING ABOUT PAYOFF ODDS

The payoffs, or *house odds*, for casino games are usually displayed on or at the betting table. A win on 20-to-1 odds should return 20 times the amount of the bet, *plus the bet itself.* So winning a $5 bet at 20 to 1 earns $100 in winnings, which, along with the $5 bet, combines for a total payoff of $105.

Some casinos count on the bettor to overlook this and take winnings of only $100. This might be listed as a payoff of 20 *for* 1, but it is technically only 19 *to* 1. Depending on the house rules, 20 *for* 1 may or may not mean 20 *to* 1; it is worth finding this out before deciding on a betting strategy.

Φ ——————————————————————— Φ

MATHEMATICAL EXPECTATION

As much as any gambler would like to know what will happen in the short run, the most he can be certain of are long-run prospects, which, when it comes to casino games, may be dimmer than he thinks.

1. The long-run prospects for any casino game are called the *expectation.* In any game of chance, the average amount a player can expect to win or lose on one play is called the *expected winnings,* or simply the *expectation.* For any bet, the expectation can be found by multiplying the probability of each possible outcome by its payoff, and then adding these results. If the expectation is positive, the player will win in the long run; if it is negative, he will lose.

• In a simple coin toss there are two possible outcomes, each with probability ½. Assume a $1 bet is placed on heads. If heads comes up, the player wins $1; if tails comes up, the player loses $1.

	Head	Tail
Probability	½	½
Payoff	$1	−$1

To find the expectation, multiply each payoff by its probability and sum the results.

$$\text{expectation} = \left(\frac{1}{2}\right) \times (\$1) + \left(\frac{1}{2}\right) \times (-\$1) = 0$$

2. An equivalent way to calculate the expectation, shown in this example, uses the house odds and the true odds.

• In American roulette, winning a bet on a single number pays 35 to 1, although the probability is only 1 in 38. To find the expectation, multiply the first number in the house odds by the probability of success, and the second number of the house odds by the probability of failure. Then subtract the second value from the first. When playing a single number in American roulette,

$$\text{expectation} = 35 \times \left(\frac{1}{38}\right) - 1 \times \left(\frac{37}{38}\right) = \frac{-2}{38} = -0.0526$$

Thus you can expect to average a 5.26 cent loss per dollar bet.

3. If a player bets the same amount repeatedly on a particular outcome, the expectation is the total gain or loss averaged over the number of plays (that is, the net gain or loss divided by the number of bets made).

• In the game of keno, a *one-spot card* may be purchased for $1. It allows the player to choose one number from 1 to 80. A dealer then chooses twenty numbers at random. If the player's number is among those chosen, she is paid $3.20.

• There are two ways to calculate the expectation. First imagine that you buy 80 one-spot tickets for $80, which allows you to play every number from 1 to 80. Twenty of the cards will be winners, resulting in a payout of 20 × $3.20 = $64. The net loss is $80 − 64 = $16 in 80 plays, for an average of ¹⁶/₈₀ = 20%. That is, you lose twenty cents for each dollar played. This 20-cent-per-dollar difference is called the *house advantage,* or the *house edge.*

• Another way to calculate the expectation is to imagine playing one number, which will cost you $1. There are two possible outcomes: (1) win and collect $3.20, for a net profit of $2.20, or (2) lose your dollar. The first has a probability of $^{20}\!/_{80}$, or 0.25; the second has a probability of 0.75. By the formula given above:

expectation = $(0.25) \times (+\$2.20) + (0.75) \times (-1) = -0.20$

This means you should expect to lose 20 cents per dollar bet. Again, the house advantage is 20%.

4. Any game in which the expectation is 0—which means the bettor wins as much as he or she loses—is called a *fair game*. In a fair game, gains are offset by losses. But there are no fair casino games. In craps, roulette, and blackjack, as well as lottery games, the expected winnings on all bets are negative; otherwise there would be no house advantage.

Gambling can be thought of as engaging entertainment (moral issues aside), which becomes all the more entertaining for the player who understands the odds going in and has a sound betting strategy. The soundest strategies are those that try to exploit the most favorable expectations. In the sections that follow, mathematical expectation is calculated for some popular games of chance—lotteries, craps, roulette, keno, and more. This is followed by a rundown of betting strategies or "systems" that fail to beat the odds but do offer some tactical advantages. A section entitled "Gambler's Ruin" assesses the prospects of any bettor who goes up against the financial resources of the casino, showing that even in a fair game a bettor who keeps playing is doomed to lose everything. Yet some people do win, and there are strategies that, while they do not guarantee victory, hold out the prospect of either quitting while ahead or keeping losses below a preset limit.

Casino Games and Pari-mutuel Betting ——————

1. Slot Machines Like roulette, slot machines constitute one of thousands of variations of the wheel of fortune. Others include bingo, keno, lotteries, and any other game in which numbers are selected at random. Because the mechanics of such games are either invisible or difficult to calculate, their mathematical expectation is not immediately apparent, which is why lotteries and sweepstakes are required to publish their odds.

Of all these games, the workings of the slot machine are the most inaccessible. A typical machine has three wheels, each displaying 20 symbols. By the principle of multiplication, there are $20 \times 20 \times 20 = 8000$ possible combinations. A four-wheel machine has 160,000 possible combinations. The symbols—bars, bells, plums, oranges, cherries, and lemons—are arranged on the wheels in varying frequencies that determine the odds against the winning combinations. These odds can vary from machine to machine, with payouts ranging from 95% to 50%. A machine with a high payout is described as "loose," while one with a low payout is called "tight."

Modern slot machines allow many different betting combinations—multiple coins, multiple windows, even machines linked together. Their primary disadvantage is how little the size of one's bets may vary. Unlike roulette or craps, slot machines prevent the use of any systems like the *martingale* (discussed below), in which bets are doubled after each loss. The slot machine is a perfect model for the study of the classical "gambler's ruin" problem, which is described in detail at the end of this section.

2. Lotteries There are many different lottery games and sweepstakes. What they have in common is that most are required to display the odds of winning and the payoffs. From these figures, the mathematical expectation can be calculated.

> ◆ A 50-cent ticket issued in a number-matching game indicates that 9,000 out of a million tickets sold (on average) will win $5. The expectation is the probability of winning (9,000 out of 1,000,000, or 0.009) times $5, plus the probability of losing (991,000 out of 1,000,000, or 0.991) times (−$0.50).
>
> expectation = (.009) × ($5) + (.991) × (−$0.50) = −$0.4505
>
> Thus a player can expect to lose about 45 cents on the average on each 50-cent ticket.

In general, the expectation on lotteries is very poor (usually resulting in a loss of 50 cents on the dollar) and is made even worse by the fact that taxes are taken out automatically. What offsets the poor odds, for most lottery players, is the size of the jackpots, but these too are deceptive; the largest jackpots are rarely paid in lump sums, but are instead parceled out in annual installments.

3. Bingo and Keno *Bingo* and *keno* are similar number-matching games. In bingo, players purchase cards on which a five-by-five grid shows 24 numbers, with a center square marked "free." The house selects numbers one at a time from the numbers 1 through 75, and players match up the numbers with their cards. The first player to get five in a row—vertically, horizontally, or diagonally—shouts out "Bingo!" and wins. The game may continue until one player has matched all of the numbers on the card. Special bets are possible, including filling the four corners or filling the eight squares around the "free" square. The house take in bingo is usually 50%, which means that only half of the money taken in by the house is paid out as prizes. Because bingo is frequently used for charitable fund-raising, and also because it is a social occasion at which few people lose very much money, it is not considered objectionable and in fact is often played in churches.

Keno is similar to bingo in principle but adapted for casino play. In keno, players purchase cards featuring the numbers 1 through 80 and select anywhere from one to 15 numbers. The object is to match any of 20 numbers that are then chosen at random. The number of matches determines the payoffs. For example, the most popular card, the *10-spot ticket*, allows the player to select 10 numbers, with the payoffs given below.

5 matches pay 2 for 1.
6 matches pay 18 for 1.
7 matches pay 180 for 1.
8 matches pay 1300 for 1.
9 matches pay 2600 for 1.
10 matches pay 10,000 for 1.

The probability of getting r matches when the player selects n numbers is given by a formula that involves combinations. Recall that "n choose r" means $C_{n,r}$, which is "n factorial" divided by the product of "r factorial" and "$(n - r)$ factorial." Any scientific calculator will do the computation; for more details, see page 201.

$$\text{probability of } r \text{ matches out of } n \text{ choices} = \frac{\binom{20}{r} \times \binom{80 - 20}{n - r}}{\binom{80}{n}}$$

The expectation for each of the keno betting cards can be calculated using the formula above. The case of the 5-spot card is worked out below in Table 5.5. In this case $n = 5$, and the number of matches must be checked for $r = 0$ (no matches) up to $r = 5$ (all numbers match). The card only pays off for 3, 4, or 5 matches.

Table 5.5. Probability and Expectation on a 5-Spot Keno Card

$$P(\text{no match}) = \frac{\binom{20}{0} \times \binom{60}{5}}{\binom{80}{5}} = 0.2271842 \text{ (lose \$1)} \qquad -0.2271842$$

$$P(1 \text{ match}) = \frac{\binom{20}{1} \times \binom{60}{4}}{\binom{80}{5}} = 0.4056861 \text{ (lose \$1)} \qquad -0.4056861$$

$$P(2 \text{ matches}) = \frac{\binom{20}{2} \times \binom{60}{3}}{\binom{80}{5}} = 0.2704574 \text{ (lose \$1)} \qquad -0.2704574$$

$$P(3 \text{ matches}) = \frac{\binom{20}{3} \times \binom{60}{2}}{\binom{80}{5}} = 0.0839351 \text{ (win \$2)} \qquad +0.1678701$$

$$P(4 \text{ matches}) = \frac{\binom{20}{4} \times \binom{60}{1}}{\binom{80}{5}} = 0.0120923 \text{ (win \$25)} \qquad +0.3023085$$

$$P(5 \text{ matches}) = \frac{\binom{20}{5} \times \binom{60}{0}}{\binom{80}{5}} = 0.0006449 \text{ (win \$331)} \underline{+0.21347}$$

$$\text{Sum of expectations} = \underline{\underline{-0.2196790}}$$

The probabilities sum to 1 because these are mutually exclusive events. The sum of the individual expectations is -0.2196790. Thus the house advantage is close to 22%. In keno, all of the cards from the 1-spot to the 15-spot have about the same expectation. Table 5.6 lists these expectations, up through the 10-spot, in the form of the house advantage. Each of these figures can be worked out following the pattern of the example above, which may be of little use to the average bettor but makes for an interesting exercise.

4. Chuck-a-luck *Chuck-a-luck,* a casino dice game that was once a popular pastime, especially in bars, is not played much today, but it provides an illustration of three-dice probabilities and associated expectations.

In chuck-a-luck, three dice—either ordinary spotted dice or special chuck-a-luck dice bearing the symbols clubs, diamonds, hearts, spades, crowns, and anchors—are placed in an hourglass-shaped wire cage. The

Table 5.6. House Percentage on Keno Tickets

Ticket	House percentage
1-spot	20.0
2-spot	21.95
3-spot	20.91
4-spot	20.96
5-spot	21.97
6-spot	20.99
8-spot	20.51
9-spot	21.26
10-spot	20.58

cage is spun and the dice fall randomly. The number of possible outcomes is $6 \times 6 \times 6 = 216$. Players bet by choosing one of the six symbols. Payoffs, if any, depend on how many times a chosen symbol appears. Table 5.7 shows the true odds and the payoff odds.

Table 5.7. Chuck-a-luck Odds

	Probability	True odds	Payoff odds
One match	$^{75}/_{216}$	141 to 75	1 to 1
Two matches	$^{15}/_{216}$	201 to 15	2 to 1
Three matches	$^{1}/_{216}$	215 to 1	3 to 1
No matches	$^{125}/_{216}$		

The probability of a loss is 125 in every 216 games. The overall expectation for chuck-a-luck can be derived from the probabilities and the payoffs given in the table.

$$\text{expectation} = \left(\frac{75}{216}\right) \times (1) + \left(\frac{15}{216}\right) \times (2)$$
$$+ \left(\frac{1}{216}\right) \times (3) - \left(\frac{125}{216}\right) \times (1)$$
$$= -\frac{17}{216}$$
$$= -0.0787$$

Thus the bettor can expect to lose almost eight cents for each dollar bet, which makes the game far less appealing than craps, which not only has a higher expectation but provides more betting options and more action.

5. Craps *Craps* is the most popular dice game. When played well, it offers one of the most sporting propositions of all casino games: a close-to-even bet, and thus a smaller house advantage than roulette, slot machines, chuck-a-luck, or keno. But it is also the fastest game in town, which makes it a good way to win or lose a lot of money fast—and a natural game for testing out a betting system.

Craps is played with two dice. A shooter rolls the dice on a craps table, and anyone can bet on the outcome. The game works like this:

i. A total of 7 or 11 on the first roll results in a win (called a *pass*).
ii. A total of 2, 3, or 12 on the first roll produces an immediate loss. Any one of these outcomes is referred to as *craps*. (When the shooter "craps out," the dice pass to the player on his or her left.)
iii. A first roll of 4, 5, 6, 8, 9, or 10 (each of which is called a *point*) allows the player to roll again.
iv. If a point from the first roll is matched before the player rolls a 7, the player wins (passes).
v. If a 7 is rolled before a point is matched, the player loses.

The probability of a win or loss on the first roll is easy to compute. From Table 5.4:

$$\text{probability of } 7 = \frac{6}{36}$$
$$\text{probability of } 11 = \frac{2}{36}$$
$$\text{probability of 7 or 11 on first roll} = \frac{6}{36} + \frac{2}{36}$$
$$= \frac{8}{36}, \text{ or } \frac{2}{9}$$
$$\text{probability of } 2 = \frac{1}{36}$$
$$\text{probability of } 3 = \frac{2}{36}$$
$$\text{probability of } 12 = \frac{1}{36}$$
$$\text{probability of 2, 3, or 12 on first roll} = \frac{1}{36} + \frac{2}{36} + \frac{1}{36}$$
$$= \frac{4}{36}, \text{ or } \frac{1}{9}$$

The probability of rolling a *point* on the first roll is $\frac{24}{36}$, or $\frac{6}{9}$.

Winning a game of craps is called "making a pass." The probability of doing so is remarkably close to ½; it is precisely $\frac{244}{495}$, or about 0.493. Conversely, the probability of losing is $\frac{251}{495}$, or about 0.507. In craps, as in most casino games, the outcome of one play is independent of the outcome of the next. There are no guaranteed winning strategies, but the best risks are the "pass/don't pass" and "come/don't come" bets. The worst bets are individual numbers and "hard way" bets. These are explained below.

Payoff odds for craps are fairly standard from casino to casino. Table 5.8 shows the basic bets with their payoffs, probabilities, and expected winnings per play. For example, on a $1 bet of *pass*, the probability of winning is 0.492929 and the probability of losing is .507071. The payoff on a win is $1. Thus the expectation is

$$(\$1) \times (.492929) + (-\$1) \times (.507071) = -0.01414$$

This means that a player can expect to lose (on average) 1.4 cents per $1 bet. (The probabilities and expectations in Table 5.8 are rounded off.)

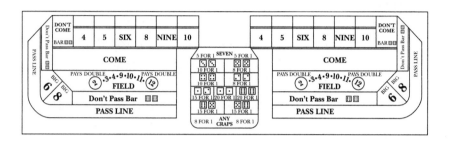

Table 5.8. Craps Table Odds and Payoffs

Bet	Payoff odds	Probability of win	Expectation
Pass	1 to 1	0.493	−0.01414
Don't pass, bar 12	1 to 1	0.493	−0.01402
Come	1 to 1	0.493	−0.01414
Don't come	1 to 1	0.493	−0.01402
Field	1 to 1	0.444	−0.056
	or 2 to 1 on 2 and 12		
Big 6 or big 8	1 to 1	0.4545	−0.0909
6 or 8 the hard way	9 to 1	0.0909	−0.0909
4 or 10 the hard way	7 to 1	0.111	−0.111
7	4 to 1	0.167	−0.167
Any craps	7 to 1	0.111	−0.111
2 or 12	29 to 1	0.028	−0.167
11	14 to 1	0.0556	−0.167
3	14 to 1	0.0556	−0.167

The bets in this table can be explained as follows:

Pass   A bet that the roller will win. This bet must be made on the first roll.

Don't pass, bar 12   A bet that the roller loses, where a first roll of 12 (on which the roller loses) is considered a tie or standoff for the bettor and there is no action on the bet. This bet must also be made on the first roll.

Come Identical to "pass," except that this bet is placed after the first roll, in which case the next roll is treated as though it were the first roll.

Don't come This bet is identical to "don't pass," but is made after the first roll. The next roll is then treated as the "come-out," or first roll.

Field A bet that the next roll produces a 2, 3, 4, 9, 10, or 12. The payoff is doubled for a 2 or 12.

Big 6 A bet that a 6 comes up before a 7 is rolled.

Big 8 A bet that an 8 comes up before a 7 is rolled.

The hard way This refers to matching dice. That is, "4 the hard way" means rolling two 2's, "6 the hard way" is two 3's, "8 the hard way" is two 4's, and "10 the hard way" is two 5's. *The easy way* refers to nonmatching dice. A bet on an even number the hard way wins if the number comes up the hard way before it comes up the easy way.

Seven A bet that the next roll is a 7.

Any craps A bet that *craps* (a 2, 3, or 12) comes up.

Numbers A bet on the specific outcome of the next roll.

Craps is a fast-paced game. While the expected losings per play are not great, the number of bets per hour (up to 130) makes it a difficult game at which to lose slowly. The safest bets are "pass", "don't pass", "come", and "don't come." The worst bets are "hard way" bets and bets on individual numbers.

6. American Roulette *Roulette* is the oldest casino game still played. It was introduced in Paris in the mid-1700s and appeared at Monte Carlo a century later. Once the most popular casino game, it still attracts many players who believe it is vulnerable to betting systems. (As Dostoevsky's Gambler found out, it is not.)

Distinct versions of the game have evolved on either side of the Atlantic Ocean. The two versions appear quite similar, but the differences have a profound effect on the odds and therefore on strategies. In both versions a wheel with numbered slots is spun while a small ball is set in motion so as to land at random in one of the slots.

In American roulette, the wheel is divided into 38 compartments numbered 1 through 36, with two more labeled 0 and 00. The numbers 1 through 36 are split irregularly but evenly between two colors—red and black—and the 0 and 00 are green. (In the illustration below, the black compartments are indicated by boldface numbers.) The ball has an equal likelihood of landing in any one of the 38 slots. Thus the probability that the ball will land in any particular slot is 1 in 38, or 37 to 1 against. At the casino table there are 13 designated bets, as well as certain combinations of these 13. With the house odds listed in Table 5.9, every bet has the same expectation of −$0.0526. This expectation, when expressed as a percent, is the house edge; thus the house edge on any bet is 5.3%, or about five cents on every dollar bet.

		0	00	
1-18	1st dozen	1	2	3
		4	5	6
EVEN		7	8	9
		10	11	12
RED	2nd dozen	13	14	15
		16	17	18
BLACK		19	20	21
		22	23	24
ODD	3rd dozen	25	26	27
		28	29	30
19-36		31	32	33
		34	35	36
		1st col.	2nd col.	3rd col.

As can be seen in Table 5.9, the true odds (or actual probability) of winning on a particular combination contrast with the house odds, which determine the payoffs the casino makes on any winning bet, the difference being the house edge.

To calculate the expectation, multiply the first number in the house odds by the probability of winning, and the second number in the house odds by the probability of losing (which is 1 minus the probability of winning), and subtract the second product from the first.

♦ If a dollar is bet on the category "1st dozen," the probability of winning is 0.3158 and the probability of losing is 0.6842. The house odds are 2 to 1. The expectation is

$$2 \times (0.3158) - 1 \times (0.6842) = -0.0526$$

Table 5.9. Odds at an American Roulette Table

Bet	True odds	House odds	Expectation
Color (red or black)	20 to 18 (.4737)	1 to 1	−0.0526
Parity (even or odd)	20 to 18 (.4737)	1 to 1	−0.0526
1–18	20 to 18 (.4737)	1 to 1	−0.0526
19–36	20 to 18 (.4737)	1 to 1	−0.0526
12 numbers or dozens (any column)	26 to 12 (.3158)	2 to 1	−0.0526
6 numbers (any 2 rows)	32 to 6 (.1579)	5 to 1	−0.0526
4 numbers (any 4-number square)	34 to 4 (.1053)	8 to 1	−0.0526
3 numbers (any row)	35 to 3 (.0789)	11 to 1	−0.0526
2 adjacent numbers	36 to 2 (.0526)	17 to 1	−0.0526
Single number	37 to 1 (.0263)	35 to 1	−0.0526

7. European Roulette In the casinos of Europe, roulette wheels have 37 numbers—1 through 36 split between 18 red and 18 black numbers, with 0 colored green. There is no 00. This makes the probabilities different from those in American roulette. The bets themselves and the house odds remain almost the same, with one notable exception. In the event that 0 comes up, all even-money bets (black/red, even/odd, low/high) are carried over to the next spin. In American roulette, such bets lose in the event of either 0 or 00.

Table 5.10 gives the true odds and house odds for European roulette, along with the traditional cosmopolitan names of the bets.

8. Blackjack *Blackjack* (also known as *twenty-one,* or *vingt-et-un*) is the most popular casino card game. When played with skill, it offers the best expectation of all casino games and is one of the few games in which a strategy can pay off. What follows is a description of the game as it is played in casinos. Variations abound, but most gambling establishments adhere to the rules given here.

Before the cards are dealt, the players place their bets within limits established by the house. Each player is then dealt one card faceup, including one to the dealer. The dealer then deals a second card faceup to each player, with his own second card facedown.

Blackjack is a game of point totals, in which numbered cards count their face value, face cards (king, queen, and jack) count as 10, and the

Table 5.10. Odds at European Roulette

Bet	True odds	House odds	Expectation
Red (rouge)	19 to 18 (.486)	1 to 1	−0.027
Black (noir)	19 to 18 (.486)	1 to 1	−0.027
Odd (impair)	19 to 18 (.486)	1 to 1	−0.027
Even (pair)	19 to 18 (.486)	1 to 1	−0.027
Low 1–18 (manque)	19 to 18 (.486)	1 to 1	−0.027
High 19–36 (passe)	19 to 18 (.486)	1 to 1	−0.027
Column of 12	25 to 12 (.324)	2 to 1	−0.027
6 numbers (sixaine)	31 to 6 (.162)	5 to 1	−0.027
4 numbers (carré)	33 to 4 (.108)	8 to 1	−0.027
3 numbers (transversal)	34 to 3 (.081)	11 to 1	−0.027
2 numbers (à cheval)	35 to 2 (.054)	17 to 1	−0.027
Single number (en plein)	36 to 1 (.027)	35 to 1	−0.027
Red-black split	28 to 9 (.243)	3 to 1	−0.027
Odd-even split	28 to 9 (.243)	3 to 1	−0.027
Split bet. 2 columns	13 to 24 (.649)	1 to 2	−0.054

ace may count as 1 or 11. The object of the game is to beat the dealer's total by getting as close as possible to 21 without going over.

After the deal the players examine their hands. Any player who has a *natural* (an ace and a 10-point card) declares it by saying "Blackjack" and wins one-and-a-half times the amount bet. (That is, if the player bets $2 and wins, he receives $3 plus the $2 bet.) If the dealer also has black-jack there is no result; the bet is returned to the player.

After blackjacks have been declared, and assuming the dealer does not have blackjack, each player may either *stand* or *draw* (by saying "hit me"). Each player plays out his or her hand before the dealer does, either reaching 21, standing at some total below 21, or busting by going over 21. When the dealer finally plays her hand, she follows the same strategy every time: she draws whenever her total is 16 or less and stands whenever it is 17 or more.

In calculating point totals, the player can choose at any time to count an ace as either 1 or 11. When the ace is counted as 11, the hand is referred to as a *soft count*. When it counts as 1, the total is a *hard count*. The dealer must count the ace as 11 anytime she reaches 17 by doing so. That is, the dealer stands on a hard or soft 17. If the dealer busts, all players left standing win an amount equal to their bet. If the dealer stands, any player with a higher total also collects even money. Only a natural 21 pays off at 3 to 2. If a player matches the dealer's total, there is no result and the bet is returned. What makes the game interesting is not the play, which is almost automatic, but the betting, which features several options:

i. *Insurance* If the dealer's facing card is an ace, the player may hedge his bet by wagering half of his original wager that the dealer has blackjack. If so, the player wins 2 to 1 on the insurance bet. If not, he loses the insurance bet but plays out the rest of his hand.

ii. *Splitting a Pair* This option turns the player's original hand of two cards into two separate hands. It may be played whenever the first two cards dealt have the same numerical value. (This includes not just two-of-a-kind but any pair of 10-point cards such as a jack and a queen, which may also be split.) The player splits a pair by matching the original bet. He then draws to both cards in the split. If he is dealt another pair, he may split that one as well and continue to split any subsequent pairs. If the player splits aces, the dealer will only deal one more card to each ace, and the payoff reduces to even money instead of the usual 3 to 2.

iii. *Take One Down for Double (Double Down)* In this option the player doubles his bet before the dealer hits his hand, agreeing to take only one more card. The rules may differ from house to house. Some allow doubling down on 9, 10, or 11; some on just 10 or 11; and some on any two cards.

The betting strategies for blackjack seem to vary from expert to expert. There are all kinds of systems, and before the application of computer analysis to the game it was impossible to know what to believe. Obviously, the player's move depends on three factors: (1) the value of the cards he holds, (2) the value of the dealer's exposed card, and (3) the values of the cards that have already appeared on the table and are no longer in the deck.

Factor 3 is the most difficult to keep track of, mainly because casinos use four or more decks simultaneously. At the very least, the player should try to keep track of how many 10's, jacks, queens, and kings have been played in relation to how much of the deck has been played. This will determine when it is wise to take out insurance, for example.

There are a few things on which all of the experts agree: Always split aces and 8s. Always double down on an 11-count. Always take out insurance whenever the dealer shows an ace and you know that the deck is heavy with 10-count cards (that is, when the ratio is more than one-third).

Table 5.11 summarizes these strategies (except insurance). It shows when to draw, when to stand, when to split, and when to double down, based on the player's count and the dealer's up-card. It takes the subjective judgments of the experts off the table, leaving it to a computer. The table was generated by Richard A. Epstein, and is reprinted from his *Theory of Gambling and Statistical Logic*, rev. ed. (Academic Press, 1995).

Table 5.11. Blackjack Strategies

Player's Hard-Count Total	Dealer's up-card									
	2	3	4	5	6	7	8	9	10	A
17										
16						x	x	x	x	x
15						x	x	x	x	x
14						x	x	x	x^f	x
13	a					x	x	x	x	x
12	x	x^b	c			x	x	x	x	x
11	D	D	D	D	D	D	D	D	D	D
10	D	D	D	D	D	D	D	D	x	x
9	D	D	D	D	D	x	x	x	x	x
8	x	x	x	x^d	D^e	x	x	x	x	x
7	x	x	x	x	x	x	x	x	x	x

Player's Soft-Count Total	2	3	4	5	6	7	8	9	10	A
19										
18		D	D	D	D			x	x	
17	D	D	D	D	D	x	x	x	x	x
16	x	x	D	D	D	x	x	x	x	x
15	x	x	D	D	D	x	x	x	x	x
14	x	x	D	D	D	x	x	x	x	x
13	x	x	D	D	D	x	x	x	x	x

Player's Pair	2	3	4	5	6	7	8	9	10	A
2's	S	S	S	S	S	S				
3's	S	S	S	S	S	S				
4's				S						
5's										
6's	S	S	S	S	S	S				
7's	S	S	S	S	S	S	S			
8's	S	S	S	S	S	S	S	S	S	S
9's	S	S	S	S	S		S	S		
10's										
A's	S	S	S	S	S	S	S	S	S	S

Key:

x = *draw;* D = *double down;* S = *split;* no symbol = *stand*

[a]Draw with 3 and 10.

[b]Stand with 3 and 9, 4 and 8, or 5 and 7.

[c]Draw with 2 and 10.

[d]Double down with 3 and 5.

[e]Do not double down with 2 and 6.

[f]Stand with 7 and 7.

9. Baccarat *Baccarat,* like its first cousin *chemin de fer,* is a European casino card game in which players seek to make a total of 9 on a two-card deal. Failing that, the object is to get as close to 9, or to a number ending in 9, as possible on subsequent draws. The rules are too complicated to explain here. What matters to any bettor is that the player has no choice in how he plays the game. There is no skill involved, and the odds do not change. Of all casino games involving no skill and pure luck (which includes everything but poker and blackjack), it has the smallest house advantage and thus represents the closest thing to an even-money bet. The house advantage is just 1.36% (compared with 5.26% for American roulette and 1.41% for the even-money bets in craps).

Because it has a higher minimum bet than most other games, baccarat is not a good candidate for a martingale betting system, in which bets are doubled after each loss (see below). It does, however, present a good testing ground for the reverse martingale, in which bets are doubled after each win. None of these systems will overcome the house advantage, however.

BETTING SYSTEMS

All betting systems try to beat the house advantage by varying the timing or the size of bets. While none of them (except for card-counting at blackjack) will overcome the house advantage, each offers a different satisfaction to the bettor, which, if it does not result in actual winnings, will at least make the game more interesting.

But before considering any of the betting systems described below, a bettor should know something about the probabilities of runs of losses. Most systems are based on even-money bets—that is, wagers in which the probability is close to ½. This includes pass/don't pass in craps; red-white, high-low, or even-odd in roulette; and games such as blackjack and baccarat. A coin toss, which *is* an even-money bet, is a good model for these games because it represents the best-case scenario in terms of odds. But it can also indicate the likelihood of a worst-case scenario, in which a run of bad luck wipes you out.

If the game is heads or tails, how often should you expect a run of three heads, or six heads, or nine heads? Nine heads is not as unlikely as it may sound. To see why, it is important to distinguish between nine heads in nine tosses and nine straight heads occurring *somewhere* in a long series of tosses. If the bet were nine heads in nine tosses, the probability would be low—$\frac{1}{512}$, or about 0.002. But if the coin were flipped 1000 times, is it unlikely that you would encounter a run of 9 heads in a row, somewhere in those 1000 flips? No.

For a large number of plays of a 50-50 game, such as tossing a coin, the approximate number of runs of heads (or tails) is given by this formula:

$$\text{expected number of runs} \approx \frac{n}{2^r + 1}$$

where n is the number of tosses and r is the length of a run.

For example, in 1000 coin tosses ($n = 1000$), the expected number of runs of 5 heads ($r = 5$) is given by:

$$\frac{1000}{2^{5+1}} = \frac{1000}{2^6} = \frac{100}{64} \approx 16$$

This means that in 1000 plays you can expect 5 heads in a row to occur almost 16 times. The same will be true for tails. Thus not only is a run of 5 heads not unusual, it is to be expected at some point in a series of bets. Table 5.12 shows the approximate expected number of runs of different lengths in extended coin-toss trials. The numbers are rounded off, but they give an idea of how many runs to expect in a long game.

Table 5.12. Expected Runs of Different Lengths in a Coin-Toss Game

Length of run	Expected number of occurrences in		
	1000 trials	*10,000 trials*	*1,000,000 trials*
3	62	625	62,500
4	31	312	31,250
5	16	156	15,625
6	8	78	7,810
7	4	39	3,900
8	2	20	1,950
9	0.967	10	975
10	0.483	5	490

The crucial numbers are 7, 8, and 9, because they represent the number of consecutive losses that will either wipe out most players who are doubling up after losses or bring them up against the maximum betting limit of most casinos (see below) and thus prevent them from adhering to the system. Every betting system has to contend with the prospect of runs of bad luck. What most of them do is to guarantee a win provided the player does not encounter seven or eight straight losses. Only when the small chance of a bad run is overlooked can the system player claim to have a foolproof way to win.

1. The D'Alembert System In light of Table 5.12, you might think it would be unnecessary to debunk the idea that an outcome may become "overdue" according to the law of averages. Yet this idea is the basis of many popular systems that propose when to bet and what to bet. The idea that when you observe a streak (for example, five consecutive reds at the roulette table) you should bet against it (black) because the alternative is somehow overdue is not a system so much as a misplaced faith.

It is possible to base a mathematical system on such a misplaced faith, which is exactly what the D'Alembert system does. The D'Alembert player selects an even-money bet (such as black-red in roulette), chooses one of the outcomes (say red), and keeps playing it. Following each loss, he adds one chip to the previous bet; following a win, he deducts one chip from the previous bet. The fallacy of the system becomes apparent when you consider that black and red are, in terms of probability, the same bet. If the system works for you, then someone who plays the same system but chooses the opposite outcome (black) would be at a disadvantage. That is, if the D'Alembert system works for player A, it must work against player B. This reduces it to the level of pure dumb luck.

2. Martingale Baron Rothschild supposedly once remarked to Monsieur Blanc, the founder of the Casino at Monte Carlo, "Take off your limit and I will play with you as long as you like." What Rothschild had discovered was the *martingale,* a system based on the simple idea that you should double your bet after every loss, and return to the opening bet after each win. The system is designed for even-money bets, and it is easy to show that the player can usually quit while ahead, provided he can leave the table after a win.

The system was invented for roulette, but it can be applied to any even-money wager. If the player bets one chip each time, after x wins he will be ahead by x chips. This is shown in Table 5.13, in which a random string of wins and losses produces one possible martingale outcome. Note how after each win the net gain equals the cumulative number of wins.

Table 5.13. A Martingale

Bet no.	Amount bet	Outcome	Net
1	1	win	+1
2	1	lose	0
3	2	lose	−2
4	4	lose	−6
5	8	win	+2
6	1	lose	+1
7	2	win	+3
8	1	win	+4

The catch is the house betting limit. Some casinos maintain a 500-to-1 ratio of high to low bets, and most have an even lower limit. Such limits were instituted specifically to thwart martingale players. This means that if a table has a $1 minimum bet, the maximum might be as low as $200, and never higher than $500. Thus the martingale player cannot keep doubling, since the limit kicks in after seven or sometimes nine consecutive losses.

Here is a succession of martingale bets that assumes a loss each time. The previous bet is always doubled.

1st	2nd	3rd	4th	5th	6th	7th	8th	9th
1	2	4	8	16	32	64	128	256

The ninth bet would not be possible in some casinos. The tenth bet, which assumes nine straight losses, would have to be $512, and is beyond every house limit. If the limit were lifted, someone with unlimited funds, like Baron Rothschild, could continue to play as long as he liked and walk out the winner. For many high rollers it is lifted, but not for the average patron.

The martingale requires resoluteness. The bettor cannot flinch in the face of a bad run, and he must have enough money to withstand such a run. There is always a chance of a run that will ruin the bettor, but that chance is relatively small. So while the expectation remains negative, the martingale bettor can usually walk away with something if he or she is willing to accept the small chance of losing everything.

3. Reverse Martingale In the reverse martingale, the bet remains constant after each loss and is doubled after each win. This keeps the bets from growing too large, unless the player keeps winning, in which case the money comes from the table and not from his pocket. This system is designed for games with a high minimum bet. While the martingale carries a high probability of modest profits, the reverse martingale carries a small chance of a spectacular win and a good chance of many small losses. It appeals to players who believe in hot streaks.

4. Great Martingale In a standard martingale, the bet is doubled after each loss. In a *great martingale* the bet is doubled and then increased by 1 after each loss. Thus the progression of bets during a series of losses would be: 3, 7, 15, 31, 63, and so on. In each case, the bet after a loss is 1 less than a power of 2. Table 5.14 shows the result of a great martingale system that follows the same sequence of wins and losses used in Table 5.13. It shows that the player stands to gain more with a great martingale than a standard martingale.

Table 5.14. A Great Martingale

Bet no.	Amount bet	Outcome	Net
1	1	win	+1
2	1	lose	0
3	3	lose	−3
4	7	lose	−10
5	15	win	+5
6	1	lose	+4
7	3	win	+7
8	1	win	+8

The downside of the great martingale is that it is more vulnerable to runs of bad luck; the great martingale player cannot withstand as long a losing streak as the martingale player can. Seven consecutive losses (which can be expected to happen 4 times in 1000 plays) will bring the player to the 200-to-1 house limit.

5. The Labouchère or Cancellation System In the *Labouchère system*, the bettor writes down a list of arbitrary numbers in the hope of winning the sum of the numbers. Like a martingale, this system is designed for even money bets. The amount of the first bet is the sum of the first and last numbers on the list. If this results in a win, these numbers are crossed off and the bettor proceeds with the new first and last numbers. In the event of a loss, nothing is crossed off; the amount lost (the sum of the pair) is written at the end of the list, and the process continues. When the bettor makes it through the entire list and all of the numbers are canceled, he will have won an amount equal to the sum of the numbers in the original list.

For example, if the list consisted of the numbers 1, 3, 5, 6, 8, and 9, the first bet would be 1 + 9 = $10. In the event of a win, the bettor crosses off the 1 and 9 and proceeds to bet 3 + 8 = $11. If this results in a loss, the bettor writes 11 at the end of the list and continues. The next bet would be 3 + 11 = $14.

What makes the cancellation system appealing is that, while the number of wins and losses should balance, each win crosses two numbers off the list but each loss adds only one. This seems to guarantee that the bettor will make it through the list and win the sum of the original numbers. The disadvantage of this system is the limit it places on winnings, which contrasts with the mounting size of bets during a run of losses. Like the martingale, the Labouchère system can run up against a house limit. More likely, the player will become discouraged in the face of inevitable runs of losses. It can take a long time to work through a list, and an impatient bettor might not make it through. Also, like the martingale, cancellation trades off a high probability of a small gain against a low probability of a big loss. When these results are weighed against each other mathematically, the expectation is still negative.

6. The 1-3-2-6 System The *1-3-2-6 system* is a variation on the reverse martingale. The player bets $1 (or one chip) on the first bet. If he wins, he bets $3; if he wins again, he bets $2; if he wins again, he bets $6. If at any point he loses, he starts over again.

At first glance, this system appears to be a sure thing. The reason is that it presents only five possibilities, and three of the five result in no loss. These are the possibilities:

First bet ($1)	Win: Go up 1, go to second bet.
	Lose: Go down 1, start again.
Second bet ($3)	Win: Go up 4, go to third bet.
	Lose: Go down 2, start again.
Third bet ($2)	Win: Go up 6, go to third bet.
	Lose: Stay up 2, start again.
Fourth bet ($6)	Win: Go up 12, start again.
	Lose: Stay even, start again.

It appears that the losing positions can occur only on the first and second bets, and that the third and fourth bets leave you even or ahead. What this fails to notice is the lower probability of getting to the third or fourth bet. The result is that the expectation is still negative.

7. There is one other system worth mentioning. It can only be used at roulette, and it seems to take advantage of an oddity of the betting layout.

In American roulette, the third column on the table (see page 221) has 8 red and 4 black numbers, and it pays off at 2 to 1. The strategy is to bet two chips on each spin: one chip at even money on "black," and one on the 2-for-1 third column. The reasoning runs this way: in 38 plays, on average, 0 and 00 should each appear once, red comes up 18 times (with 8 of these occurring in the third column), and black appears 18 times (losing for the 12 times it appears outside of the third column, but winning on the other bet). It would seem that in 38 plays, the losses on 0 and 00 would total 4 chips (recall that 2 chips are bet on each play); the losses on red would total 4 chips (because the 10 red numbers outside of the third column cost 20 chips, but the 8 in the third column would bring in 16 chips); and the winnings on black would total 12 chips (the winnings on black offset the losses outside of the third column, but the four cases of "black and third column" would net 12). All told, this would result in a net gain of 4 chips per 38 plays.

It sounds good, but the explanation obscures the true odds. The table below separates the two bets in order to calculate the total expectation.

Bet	Payoff	P(win)	Net	P(loss)	Net
3rd col.	2 to 1	$^{12}/_{38}$	+2	$^{26}/_{38}$	−1
Black	1 to 1	$^{18}/_{38}$	+1	$^{20}/_{38}$	−1

$$\text{expectation} = \left(\frac{12}{38}\right) \times (+2) + \left(\frac{26}{38}\right) \times (-1)$$
$$+ \left(\frac{18}{38}\right) \times (+1) + \left(\frac{20}{38}\right) \times (-1)$$
$$= -\frac{4}{38} \text{ per \$2 bet}$$
$$= -\frac{2}{38}, \text{ or } -0.0526 \text{ per \$1 bet}$$

The result is that this strategy has the same negative expectation as every other bet at the roulette table. So where is the fallacy in this system's reasoning? The analysis above said that the 8 red numbers in the third column bring in 16 chips. But this accounts only for the winnings on the "3d column" bet, neglecting the lost "black" bet on those numbers. The additional loss of 8 chips converts the purported net gain of 4 chips into a net loss of 4 chips, confirming the negative expectation.

GAMBLER'S RUIN

The *gambler's ruin problem* (referred to by mathematicians as the *classical ruin problem*) pits an imaginary gambler against an imaginary casino and calculates the odds that, playing only even-money bets, the gambler either goes broke or "breaks the bank." The surprising result is that even if a casino offered a "fair" game, the probability of the gambler's ruin is very high because it depends only on how much money he has compared to the resources of the house.

The analysis of the gambler's ruin problem is complicated, but a stripped-down example will produce the correct numbers.

A gambler walks into a casino with x dollars in his pocket. The house has resources of $Y–x$ dollars, so there is a total of Y dollars at stake. Let's assume that the game is a coin toss and that the house pays even money—in other words, a fair game, in which the player's losses should offset his wins in the long run. The gambler always bets the same amount. How, then, can such a game result in the gambler's ruin?

An extremely simplified scenario shows how. Imagine that the gambler has $1 and the house has $2, so there is a $3 pool at stake. The bet is $1 on each play. The chart below shows every possible outcome through the first 10 plays. If the player dips down to $0, the game is over and the house has won; but if the player's take reaches $3 (the entire pool), he has broken the bank.

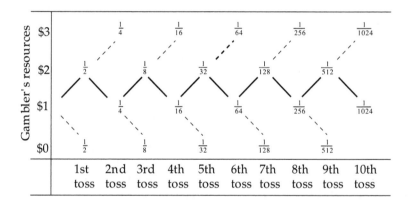

The solid zigzagging line through the middle of the figure above represents the only way the game can continue. It shows that the probability of the game lasting 10 plays is $\frac{1}{1024}$. As time goes on this probability rapidly diminishes, to the point where the chance of the game lasting 20 plays is the same as the probability of tossing 20 heads in a row—about one in a million. The probability of the game lasting 50 plays is the same as that of 50 heads in a row—about one in a quadrillion.

The bottom row of the figure shows the ways the gambler can be ruined: on the first toss, on the third, on the fifth, seventh, or ninth. The top row shows how he can break the bank: on the second, fourth, sixth, eighth, or tenth toss. These are all mutually exclusive events; the occurrence of any of them precludes the possibility of the others. Thus, to find the probability of ruin through ten plays, add the probabilities from the bottom row. To find the probability of breaking the bank through ten plays, add the probabilities in the top row. What remains will be the probability that the game has not been decided after 10 plays.

$$\text{probability of ruin} = \frac{1}{2} + \frac{1}{8} + \frac{1}{32} + \frac{1}{128} + \frac{1}{512} = \frac{341}{512}$$

$$\text{probability of success} = \frac{1}{4} + \frac{1}{16} + \frac{1}{64} + \frac{1}{256} + \frac{1}{1024} = \frac{341}{1024}$$

$$\text{probability that the game goes on} = \frac{1}{1024}$$

What can we conclude? The probability of ruin is twice the probability of success. This is not accidental, but is instead directly related to the fact that the house began with twice as much money as the gambler. The probability that the game goes on is minuscule and will continue to shrink rapidly.

In general, in a fair game,

$$\text{probability of gambler's ruin} = 1 - \frac{\text{gambler's stake}}{\text{total stake}} = \frac{\text{house's stake}}{\text{total stake}}$$

This is the *eventual probability* of the gambler's ruin, assuming the game goes on indefinitely. It says that the gambler's odds of eventual ruin are proportional to the house's resources compared to his own. With more money available to house and gambler, the game has an increased chance of continuing. But when all of the possible strings of wins and losses are considered, the higher proportion leads to the gambler's ruin, and that proportion is determined by the ratio of money held by the gambler and the house.

If the game is not fair (which is the case with all casino games), the probability of ruin increases. It is given by the formula

$$P(\text{ruin}) = \frac{\left(\dfrac{q}{p}\right)^c - \left(\dfrac{q}{p}\right)^d}{\left(\dfrac{q}{p}\right)^c - 1}$$

where p is the probability of a win on one play of the game, q is the probability of a loss, d is the player's stake, and c is the total stake.

• In a game of craps, the player comes to the table with $10 and makes $1 bets on "pass," which has a probability of about 0.49. Thus

$p = .49$, $q = 1 - .49 = .51$, and $q/p = .51/.49 = 1.041$. The house has $90, making a total stake of $100. The probability of the player's ruin is

$$P(\text{ruin}) = \frac{(1.041)^{100} - (1.041)^{10}}{(1.041)^{100} - 1} = 0.991$$

The expected duration of the game is $c \times d = 1000$ plays.

HORSE RACING

The system of betting used at all thoroughbred, quarter-horse, harness, and dog racing tracks in North America is called the *pari-mutuel* system. (This is also the system used for jai alai betting.) "Pari-mutuel" is a French term that refers to the idea of shared betting. In this system, the total amount wagered on each race is divided among the winners after a house share (or *take*) has been deducted.

1. In the pari-mutuel system, a wager may be placed on any horse to *win* (finish first), *place* (finish first or second), or *show* (finish first, second, or third). The amounts of money wagered in each category are totaled to form three separate pools. Other bets (specialty bets such as the *trifecta*, *quinella*, or *daily double*) are combined in a separate pool. A machine called a *totalizer* calculates the odds on each horse to win based on the amounts bet, and also sets payoff odds that reflect a certain percentage (usually 15–20%) taken out of the pool by the track and the state. The amounts bet (to win) and the odds against each horse are displayed on a *tote board*, which is periodically updated while the betting windows are open.

2. A horse that is a "15-to-1 shot" would return a profit of $15 for every dollar bet. Because the cheapest ticket is a $2 ticket, a single ticket at 15 to 1 would earn $30 in profit (plus the original $2). A winning ticket on a horse that won against odds of 2 to 5 would pay $2 on a $5 bet, or $4 on a $10 bet. A $4 profit on a $10 bet is equivalent to an 80-cent profit on a $2 bet.

The amount bet plus the profit is referred to as the *extension* or *payoff*—that is, the amount that is actually paid out when a winning ticket is presented at the ticket window. The extension is, in a sense, the money on the table; all of it goes either to the bettor or to the house.

3. If $20,000 is bet on a race, the house take would be 15% of $20,000, or $3,000. This leaves $17,000 to be split among the winners. However, the track automatically rounds the payoff on a $2 bet down to the nearest dime. This further reduces the $17,000 winners' pool by an amount called the *breakage*. The breakage, which adds to the house take, is not listed on the tote board.

4. Table 5.15 shows the amount bet to win, place, and show on an eight-horse field. Table 5.16 shows the odds and payoffs to win as they would be calculated by a totalizer. The totalizer would generate two more tables for the payoffs to place and show, but Table 5.16, without the breakage, is all that the bettors would see on the tote board. The odds will change as the betting progresses, with a final tally showing at post time.

The odds are based on the amounts bet on each horse to win. For this race this total is $77,000. The track skims off 15%, which leaves

$$\$77,000 - 15\% = 77,000 - 11,550 = \$65,450$$

to be distributed as winnings.

Table 5.15. Amounts Bet on an Eight-Horse Field

Post	Horse	Amount bet to		
		Win	Place	Show
1	Pacemaker	3,500	2,000	2,500
2	Litmus Test	10,000	8,000	8,000
3	Storm Window	5,000	5,000	3,500
4	Parboiled	10,000	6,000	7,000
5	Kreb's Cycle	12,000	7,000	10,000
6	Glue Pot	1,500	1,000	500
7	Junk Mail	9,500	5,000	4,500
8	Rutabaga	25,500	18,000	20,000
	Totals	77,000	52,000	56,000

Table 5.16. Odds and Payoffs on an Eight-Horse Field

Post	Horse	Odds given	Payoff	Total payoff	Breakage
1	Pacemaker	17 to 1	$37.40	65,450	0
2	Litmus Test	5 to 1	$13.00	65,000	450
3	Storm Window	12 to 1	$26.00	65,000	450
4	Parboiled	5 to 1	$13.00	65,000	450
5	Kreb's Cycle	4 to 1	$10.90	65,400	50
6	Glue Pot	43 to 1	$87.20	65,400	50
7	Junk Mail	6 to 1	$13.70	65,075	375
8	Rutabaga	3 to 2	$5.10	65,025	425

(Note that payoff + breakage = amount bet − 15%.)

5. To calculate the odds on Pacemaker, subtract the amount bet to win—$3500—from the total stake of $65,450, which leaves $61,950 to be divided as profit among the winners if Pacemaker finishes first. Then divide by the total amount bet to derive the profit per dollar bet.

$$\$61,950 \div 3500 = \$17.70$$

In this case the totalizer rounds 17.70 for 1 to 17-to-1 odds. The actual payoff on a $2 bet (the standard bet) would be

$$2 \times \$17.70 = \$35.40$$

Add to this the value of the bet itself—$2—to get the listed payoff of $37.40. Because this figure does not need to be rounded (payoffs are rounded to the nearest dime), there is no breakage.

Here is the same calculation for Junk Mail:

Stake: (amount bet to win) − 15% = $77,000 − 11,550 = $65,450
Profit pool: (stake) − (amount bet) = $65,450 − 9500 = $55,950
Profit per dollar bet: $55,950 ÷ 9500 = $5.89 (rounded to 6-to-1 odds)
Profit per $2 bet: 2 × $5.89 = $11.78 (rounded down to $11.70)
Payoff per $2 bet: $11.70 + 2 = $13.70
Total number of $2 bets: 9500 ÷ 2 = 4750.
Total payoff: 4750 bets × $13.70 per bet = $65,075
Breakage: (stake) − (total payoff) = $65,450 − 65,075 = $375

6. Assume the order of finish is: (1) Parboiled, (2) Rutabaga, and (3) Glue Pot. The first-place payoff on Parboiled has already been calculated. All holders of bets on Parboiled and Rutabaga to place would win, with the amounts calculated as follows:

Total amount bet to place: $52,000
Stake: (amount bet) − 15% = $52,000 − 7800 = $44,200
(A total of $6000 + 18,000 = $24,000 was bet on the two horses to place.)
Payoff: $44,200 − 24,000 = $20,200, to be divided among the winners. (Divide this by 2 to get $10,100 for each horse.)

In the table below, the amount bet is added to the amount won in order to derive the extension. This is divided by the number of bets to give the payoff per $2 ticket. This number is rounded down to the nearest dime.

	Parboiled	Rutabaga
Amount bet	$6,000 (3000 bets)	$18,000 (9000 bets)
Amount won	10,100	10,100
Theoretical payoff	16,100 ÷ 3000 = 5.366	28,100 ÷ 9000 = 3.122
Actual payoff	$5.30	$3.10
Total payoff	3000 × $5.30 = $15,900	9000 × $3.10 = $27,900
Breakage	3000 × $0.066 = $200	9000 × $0.022 = $200

7. The show bets totaled $56,000. Taking 15% off the top leaves $47,600, of which $27,500 was the amount bet by the winners on the three horses to show. The difference ($47,600 − 27,500 = $20,100) represents profit to the winning bettors, and it is divided three ways. The result is $6700, which is split among the show ticket holders for Parboiled, Rutabaga, and Glue Pot in proportion to the number of winning show tickets on each horse.

	Parboiled	Rutabaga	Glue Pot
Amount bet	$7,000 (3,500 bets)	$20,000 (10,000 bets)	$500 (250 bets)
Amount won	6,700	6,700	6,700
Theoretical payoff	13,700 ÷ 3500 = 3.9143	26,700 ÷ 10,000 = 2.67	7,200 ÷ 250 = 28.80
Actual payoff	$3.90	$2.60	$28.80
Total payoff	3500 × $3.90 = $13,650	10,000 × $2.60 = $26,000	250 × $28.80 = $7200
Breakage	3500 × $0.0143 = $50	10,000 × $0.07 = $700	0

8. State laws mandate a minimum payoff on any winning ticket, usually $2.10 or $2.20 on a $2.00 ticket. Sometimes, when too much money is bet on one horse, the pool of place and show money may not be enough to cover the payoff on winning tickets, and the track must make up the difference.

Statistics

Statistics is the science of manipulating raw data into usable information—of organizing and summarizing polls, samples, and measurements that generate lists of numbers. The first step in organizing such data is to organize it into a *distribution*, which may then be evaluated to detect patterns or tendencies. When two data sets are compared to each other, a statistician may notice a relationship, or *correlation*, that could hint at unexpected patterns of cause and effect. The subject of statistics is immense, and its products reflect many aspects of everyday life in the form of opinion polls, astounding numerical facts, and profiles of the average American, often becoming the basis for public policy initiatives and legislation. The American media today both create and feed a craving to know what everyone else is thinking, doing, buying, watching, eating, and wearing. In the midst of all these statistics, it might be useful to look at how such "facts" are determined.

DISTRIBUTIONS

A distribution is essentially a set of measurements. Any list of numbers generated from a single experiment or test constitutes a distribution. Exam scores, per-capita incomes, heights, weights, and IQs of a given sample population all constitute distributions. A statistician's primary concern when confronted with any distribution is whether the sample is big enough and representative enough—that is, whether it adequately reflects the entire population or is instead biased. The more measurements a distribution contains and the more random they are, the more accurately the test summarizes the situation. But bias is difficult to eliminate. Political pollsters try to assess the mood of an entire country by polling 1000 people at a time, yet they often fail to predict the way the electorate will vote. The Nielson rating system uses a cross-section of American families to reveal the television viewing habits of the entire country, yet the system is far from perfect. Every sample has error and bias, and researchers constantly struggle to achieve reliability. The statistician's task is easier. He or she is primarily concerned with analyzing the numbers. The statistician's job is to eliminate opinions from the analysis of data.

When a distribution consists of a very large list of numbers, it is useful to have some way of summarizing it. What is the range of numbers involved? What is its middle value? How are the numbers spread? Do they cluster around a middle value or are they spread over a wide range with very little clustering? Such questions generate a set of numbers called *statistics*. These include the *mean, median, mode, spread, variance,* and *standard deviation*.

This section is about the most basic statistics and how they can be used to convey the essential trends or results contained in a distribution. It begins with a brief overview of distributions and ways they are typically represented.

1. Bar Graphs Distributions begin as lists of numbers. Once the numbers are organized in the form of a distribution, a few generalizations can be made. One way to visualize a distribution is to represent it with a bar graph or a pie chart. In a bar graph, the length or height of each bar represents the frequency of a value in a list of values. Thus a bar graph is a graph of a *frequency distribution*.

♦ Assume that a test given to ten students resulted in the following scores: 100, 90, 90, 80, 80, 80, 80, 70, 70, 60. These results can be summarized in a frequency distribution table.

Score:	100	90	80	70	60
Frequency:	1	2	4	2	1

A bar graph for this distribution would consist of a bar for each test score, whose height would be proportional to the frequency of that score.

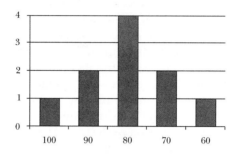

2. Pie Charts A pie chart is another way of representing the same kind of distribution. To create a pie chart, first change each frequency to a percent, as shown here.

Score	100	90	80	70	60
Frequency	1	2	4	2	1
Percent	10%	20%	40%	20%	10%

The corresponding pie chart represents the frequency of each score as a proportional slice of the pie.

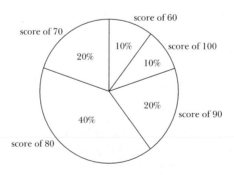

There are other ways to summarize a data set. Graphs and charts have the advantage of immediate visual impact. They can quickly summarize a key idea or trend contained in a data set. But they do not so clearly convey other summary information. The pie chart in particular is useful in showing the relative size of categories but not very good at showing how the data looks within categories or how it looks as a whole. These issues can be addressed by supplementing the chart or graph with a few useful numbers: (1) the *range*, which describes the difference between the highest and the lowest values; (2) the *average*, which describes the "middle value"; and (3) the *standard deviation*, which describes how much the distribution clusters around its middle value.

THE RANGE OF A DISTRIBUTION

1. The *range* of the set of numbers 3, 25, 27, 27, 28, 29, 33, 53 is the difference between the high and low values: $53 - 3 = 50$. This is also called the *spread*. It is simply an interval within which all of the other values fall.

2. Whenever the low and high terms of a distribution are out of scale with the middle terms, it may be more informative to define a range that eliminates the extreme values. This is sometimes appropriate with test scores, where one unusually good or extremely bad grade is not indicative of the performance of the group as a whole.

There are several schemes for showing a spread that eliminates numbers or scores at the low and high end. Most take a certain percentage off the top and the bottom. One way to do this is to use *percentiles*.

3. Percentiles As discussed in Chapter 1, a *percentile* is one of 100 equal divisions of a distribution. In general, a *percentile rank* identifies a group or range of values by the percent of values that fall below it. The 50th percentile is the top half of a set of ordered numbers. If a group is divided into 10 equal groups in order of rank, each subgroup is called a *decile*. If the original group is divided into 4 subgroups, each subgroup is called a *quartile*.

One method of giving the range in a distribution is to lop off the top and bottom deciles, leaving a range consisting of the middle eight deciles. Another scheme is to lop off the top and bottom quartiles. What remains is called the *interquartile range*. The interquartile range represents the "middle" half of the distribution.

AVERAGES: MEAN, MEDIAN, AND MODE

Every distribution has several "middle" values which are referred to collectively as *averages*. Which type of average you work with depends on how you define "middle." There is a middle *value* called the *mean*, a middle *number* called the *median*, and possibly a most frequently occurring number called the *mode*. There is also a specialized average known as the *geometric mean,* which applies to lists of numbers that exhibit a particular type of growth.

1. The most common mathematical average is the *arithmetic average,* or *mean.* Given any set (or list) of numbers, the mean is found by summing all of the numbers in the list, and then dividing by the total number of numbers.

- The mean of 2, 5, 6, 7, 10, 20, 25, and 75 is

$$\frac{2 + 5 + 6 + 7 + 10 + 20 + 25 + 75}{8} = \frac{150}{8} = 18.75$$

Because it is an average value, the mean does not have to be a member of the list. Therefore the word *average* should be used with care. A statement such as "The average family has 2.75 children" raises the absurdity of three-fourths of a child. Properly stated, the statement should read, "The average number of children per family is 2.75."

2. Another type of average, one that can be more useful than the mean in a list that has some disproportionately large or small values, is called the *median.* In an ordered list of numbers, the median is the middle number; that is, if the numbers are listed in order from smallest to largest, the median is the number that falls in the middle of the list. (If the list has an even number of elements, the median would be the arithmetic average of the two middle numbers.)

- The median of 18, 24, 27, 30, 35, 42, 50 is 30.
- The median of 4, 5, 7, 10, 14, 22, 25, 30 is the average of 10 and 14, which is 12.

3. The *mode* is the number that occurs most frequently in a list of numbers. It is particularly important in any study that tracks the most popular choice in a vote or poll.

- Consider the list of numbers: 2, 3, 7, 7, 8, 12, 15, 26.

The *mean* is given by: $(2 + 3 + 7 + 7 + 8 + 12 + 15 + 26) \div 8 = 10.$
The *median* is the average of 7 and 8, which is 7.5.
The *mode* is 7.

4. Just as the word *average* is almost always understood to imply the *arithmetic average*, the word *mean* is almost always understood to imply the *arithmetic mean.* This should be distinguished from another type of mean, the *geometric mean.* The arithmetic mean is typically used with sets of numbers that are spread out somewhat evenly. A good example would be a list of daily high temperatures in a given month. By contrast, the geometric mean applies to numbers that increase or decrease rapidly because each term is a fixed multiple of the previous term. This is a growth pattern commonly seen in population studies and in investments.

The geometric mean is defined to be the nth root of the product of the *n* numbers in a list of numbers.

* The geometric sequence 1, 2, 4, 8, 16 has five terms. Its geometric mean is $\sqrt[5]{1 \cdot 2 \cdot 4 \cdot 8 \cdot 16} = \sqrt[5]{1024} = 4$. Notice that this is the same as the middle term of the sequence, which is the median in this example. This will not always be the case. The arithmetic mean of this set is $(1 + 2 + 4 + 8 + 16)/5 = 3\frac{1}{5} = 6.2$.

* Assume that the population of a city was 46,000 in 1980 and 73,000 in 1990. To estimate the population in 1985, one could find the arithmetic mean of 46,000 and 73,000, which is 59,500. But because populations tend to grow exponentially (that is, they double at fairly regular intervals, and thus tend to follow a geometric progression such as the one used in the example above), a geometric mean would be more appropriate. In this case, with only two numbers, the geometric mean would be $\sqrt{46,000 \times 73,000}$, which is about 58,000.

5. The arithmetic mean is not appropriate in certain situations, because it can be skewed by a single value that is out of proportion with the other numbers on a list. In the following example, the arithmetic mean could be appropriate, depending on what it is supposed to be showing.

* The gross annual incomes of families living on a certain street are $35,000, $37,000, $42,500, and $51,000. The mean income is

$(35,000 + 37,000 + 42,500 + 51,000)/4 = \$41,375$

If a millionaire, with a yearly income of $975,000, builds an estate at the end of the street, the mean gross annual income becomes $228,100. But this is not representative of the group. The median income, by contrast, would be $42,500, which gives a better idea of the prevailing income level, although it disguises the fact that there is a millionaire on the block. The choice of which average to use depends on the impression one wishes to convey. Consumers of statistics should bear this in mind whenever the word *average* crops up.

6. A *weighted average*, unlike an arithmetic average, assumes that each number in a set of numbers carries a predetermined weight, which is given as a percent. The sum of the weights in any weighted average amounts to 100%.

To calculate a weighted average of a set of numbers, multiply each number in the set by its weight (expressed as a decimal), then sum them up.

◆ A student's average in a college chemistry course is weighted in the following way:

Midterm exam:	25%
Final exam:	40%
Labs:	20%
Homework:	15%
Total	100%

If the student scored 75 on the midterm, 80 on the final exam, 85 on the labs, and 90 on the homework, her weighted average would be

$$75 \times (.25) + 80 \times (.40) + 85 \times (.20) + 90 \times (.15) = 81.25$$

STANDARD DEVIATION

A useful description of the spread of a distribution is provided by the *standard deviation*. The standard deviation can best be determined by using a computer program or a calculator. It may be calculated by hand if only a small set of numbers is involved, as in the simple example given below.

1. Calculation of the Standard Deviation of a Distribution of Numbers

Assume that a distribution consists of the numbers 2, 4, 7, and 11. The standard deviation is calculated by starting with the mean.

i. Calculate the mean:

$$\text{mean} = \frac{2 + 4 + 7 + 11}{4} = \frac{24}{4} = 6$$

ii. Subtract the mean from each number in the set, which results in a set of *deviations from the mean:*

$$2 - 6 = -4$$

$$4 - 6 = -2$$

$$7 - 6 = 1$$

$$11 - 6 = 5$$

iii. Square each deviation, and add up the resulting squared deviations:

$$(-4)^2 = 16$$

$$(-2)^2 = 4$$

$$(1)^2 = 1$$

$$(5)^2 = 25$$

$$16 + 4 + 1 + 25 = 46$$

iv. Divide the sum of the squared deviations by the number of elements in the set:

$$46 \div 4 = 11.5$$

This number is called the *variation.*

v. To find the standard deviation, take the square root of the variation: $\sqrt{11.5} \approx 3.39$

The standard deviation of 3.39 is an average of the deviations from the mean. In other words, it measures how the numbers in the set are spread out around the mean value.

2. Interpreting Standard Deviation The standard deviation provides a quick summary of the proportion of scores that fall a given distance from the mean. With a sample of four values, as in the example above, the standard deviation is relatively uninteresting. But with a large distribution, such as one representing income levels of the population of a city, the standard deviation is a useful way to summarize a vast quantity of data.

This is done by dividing the distribution into intervals having widths that are multiples of the standard deviation. For example, if an exam is given to 1000 students, and the mean score is 71 with a standard deviation of 5, then the group that is *one standard deviation from the mean* is the group that scored between 71 − 5 and 71 + 5, which is the interval from 66 to 76. The range of scores falling *two standard deviations from the mean* falls between 71 − 10 and 71 + 10, which is the interval from 61 to 81. *Three standard deviations from the mean* includes scores between 56 and 86, or 15 units to either side of the mean.

The general rules for interpreting standard deviation are:

> 68% of the scores fall within one standard deviation of the mean.
> 95% fall within two standard deviations of the mean.
> 99.7%, or almost all of the scores, fall within three standard deviations of the mean.

BELL CURVES

The classic bell curve is the shape of a *normal distribution*. A normal distribution is any set of numbers that is distributed evenly and symmetrically around its mean in a shape that looks much like a classic bell.

Most measures of physical characteristics—heights, shoe sizes, hat sizes, and so on—fall into normal distributions. The results of a two-dice toss, recorded over thousands of trials, will also produce a distribution of numbers which, if graphed in a bar graph, would look like a bell-shaped curve.

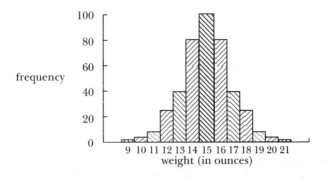

The peak of the bell falls at the mean. In fact, in any normal distribution the mean, median, and mode coincide. As many scores or measures fall above the mean as below. Because the normal distribution is the most useful distribution, it has been tabulated in Table 5.17 in a way that allows probabilities to be calculated easily.

1. Normally Distributed Data Sets A distribution is said to be *normally distributed* if its frequency bar graph can be accurately modeled by a normal, or bell-shaped, curve. Such a distribution will have a *mean* and a *standard deviation,* which can easily be calculated by machine. These numbers are always reported in any statistical summary, and can be used in conjunction with the *standard normal density table* to perform two basic tasks:

i. To examine how the scores or measurements are distributed (or spread out) around the mean. (Do they all cluster right around the mean, or are they spread out over a wide range of values?)

ii. To make predictions about how similar experiments will come out in the future. (That is, to determine the probability that in future studies, or even in individual instances, a certain result will be obtained.)

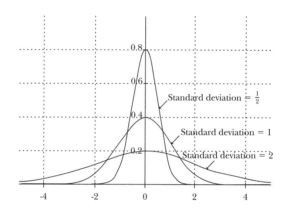

2. How to Convert Normal to Standard Normal *In a standard normal distribution, the mean is 0 and the standard deviation is 1.*

All normally distributed data sets can be converted to *standard* normal using a simple formula involving their mean and standard deviation. In other words, any data set that corresponds to a bell curve can be made to correspond to the standard bell curve, the one that is centered at zero and has the "standard" bell shape (in which the standard deviation is 1). This is done by using the mean as a point of reference. Ranges of data values are then stepped off in "standard deviations from the mean."

Data sets that are normally distributed but not *standard* normal will have different standard deviations, depending how much the data clusters around the mean. But in all of them, the following pattern will hold:

68.26% of all values will fall within 1 standard deviation of the mean.

95.44% of all values will fall within 2 standard deviations of the mean.

99.73% of all values will fall within 3 standard deviations of the mean.

♦ If a set of exam scores is normally distributed with a mean score of 72 and a standard deviation of 6, then:

The range of scores that lies 1 standard deviation from the mean is 66 to 78. This includes about 68% of all scores.

The range of scores that lies 2 standard deviations from the mean is 60 to 84. This includes about 95% of all scores.

The range of scores that lies 3 standard deviations from the mean is 54 to 90. This includes about 99% of all scores.

3. To convert normal to *standard* normal ranges, subtract the mean from each range boundary value, then divide by the standard deviation.

♦ In the example given above:

The range from 66 to 78 becomes $\dfrac{66 - 72}{6}$ to $\dfrac{78 - 72}{6}$, or -1 to 1.

The range from 60 to 84 becomes $\dfrac{60 - 72}{6}$ to $\dfrac{84 - 72}{6}$, or -2 to 2.

The range from 54 to 90 becomes $\dfrac{54 - 72}{6}$ to $\dfrac{90 - 72}{6}$, or -3 to 3.

Thus 68.26% of all values fall between 66 and 78, 95.44% of all values fall between 60 and 84, and 99.73% of all values fall between 54 and 90.

4. To find the *interquartile range* (the middle 50%) of a normally distributed data set: Multiply the standard deviation by ⅔. Construct an interval by adding this value to the mean, and subtracting it from the mean. This range constitutes the middle 50%, or interquartile range.

♦ A set of exam scores has a mean of 73 and a standard deviation of 6. Assuming the number of scores is large, and the results are normally distributed, the middle 50% can be found by adding ⅔ × 6 = 4 to the mean to get 77, and subtracting it from the mean to get 69. Thus 50% of the scores fell into the range from 69 to 77.

5. Reading a Standard Normal Density Table Table 5.17 shows what percent of a standard normal data set (or one that has been converted to standard normal) falls within certain ranges. The range boundaries are referred to as *z-values* or *z-scores*. Because the graph is the same shape on either side of 0, the table only gives z-values between 0 and 3.

Table 5.17. z-scores (Normal Distribution)

z	1 Greater than z	2 Less than z	3 Between −z and z
0.0	.5000	.5000	.0000
0.1	.4602	.5398	.0796
0.2	.4207	5793	.1586
0.3	.3821	.6179	.2358
0.4	.3446	6554	.3108
0.5	.3085	.6915	.3830
0.6	.2743	.7257	.4514
0.7	.2420	.7580	.5160
0.8	.2119	.7881	.5762
0.9	.1841	.8159	.6318
1.0	.1587	.8413	.6826
1.1	.1357	.8643	.7286
1.2	.1151	.8849	.7698
1.3	.0968	.9032	.8064
1.4	.0808	.9192	.8384
1.5	.0668	.9332	.8664
1.6	.0548	.9452	.8904
1.7	.0446	.9554	.9108
1.8	.0359	.9641	.9282
1.9	.0287	.9713	.9426
2.0	.0228	.9772	.9544
2.1	.0179	.9821	.9642
2.2	.019	.9861	.9722
2.3	.0107	.9893	.9786
2.4	.0082	.9918	.9836
2.5	.0062	.9938	.9876
2.6	.0047	.9953	.9906
2.7	.0035	.9965	.9930
2.8	.0026	.9974	.9948
2.9	.0019	.9981	.9962
3.0	.0014	.9986	.9973
3.1	.0010	.9990	.9980

For a standard normal distribution, column 1 gives the proportion of the distribution that falls above the value specified by z; column 2 gives the proportion of the distribution that falls below the given z value; and column 3 gives the proportion of the distribution that falls z units on either side of the mean.

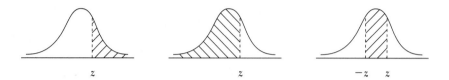

◆ The Stanford-Binet IQ test has a mean score of 100 and a standard deviation of 15. The proportion of those who score above 118 would be calculated as follows:

$$z = \frac{118 - 100}{15} = 1.2$$

Using Table 5.17, read column 1 at $z = 1.2$. It shows that .1151, or about 11½%, score above 118. This is the same as the percent that score below 82 (which is 18 points *below* the mean). Column 3 at $z = 1.2$ shows that .7698, or almost 77% of the group, fall within 18 points of the mean score (that is, between 82 and 118).

CONFIDENCE INTERVALS

Assume that a poll of 2000 people is conducted. 70%, or 1400 people, say they approve of a ban on smoking in restaurants, and 30%, or 600, say they disapprove. How reliable is the conclusion that 70% of all residents support a smoking ban?

Any conclusion based on a sample will have some error in it. This is why polls indicate a range of reliability. One poll might be accurate to within three percentage points, for example, another to within two. How much confidence can we place in such polls?

A lot has to do with how the poll is conducted. Does it truly represent a cross-section of the voting population? Is one bloc disproportionately represented? Assuming we have reason to believe the answer to the first question is yes, and the answer to the second question no, we are still left to ponder whether 2000 people, even randomly chosen people, can reliably speak for 250 million others.

This question can be addressed mathematically by using a *confidence interval*, which is based on the standard deviation. In a situation like this, where the polling results in one of two responses—yes or no—the standard deviation can be estimated by this approximation formula:

$$\text{standard deviation} = \sqrt{(\textit{number} \text{ who vote yes}) \times (\textit{percent} \text{ who vote no})}$$

In this case we have:

$$\text{standard deviation} = \sqrt{1400 \times 0.3} \approx 20$$

In any normally distributed data set, there is a 95% chance that an outcome falls within two standard deviations of the mean. Because in this poll the standard deviation represents 1% of the sample ($^{20}/_{2000}$ equals 1%), two standard deviations is equivalent to 2%. That is, we can say with 95% confidence that the results of the poll are accurate to within 2%. This is how news organizations can conduct spot polls of 2000 voters and then claim that the poll is accurate to within 2%.

Correlation

Two quantities are said to be *correlated* if there is a statistical relationship between them. Ideally, a correlation can be used to make predictions or to construct theories of cause and effect. But the fact of a correlation alone does not automatically establish cause and effect. The amount of rain that falls in Portland, Oregon, might very closely correlate with alcohol-related injuries on Florida highways, but that would not prove that one caused the other. However, when cause and effect seems probable, the researcher can design a test in which the values of one quantity are varied repeatedly in order to see if this causes variations in the effect.

For example, statistics show that fluoridation of drinking water is correlated to a decrease in tooth decay. Can it be concluded that fluoridation leads to a reduction in tooth decay? It cannot be proved by a single experiment, but several long-range studies can provide strong evidence of a causal relationship. What mathematics provides is a way of measuring the degree of correlation between two quantities.

There are three basic classifications of correlations. When two quantities are examined in isolation, they may be found to have (1) no relation whatsoever, (2) a hard-and-fast relationship, or (3) a probable relationship.

Consider an experiment in which two factors are isolated in such a way that the effect of one factor on the other can be studied. Other factors that might explain the correlation are eliminated as much as possible. The test then produces data in the form of tables of values—inputs and outputs—called the *data set,* or simply the data. If the test successfully demonstrates a correlation, it means that the graph of the data (a set of plotted points referred to as a *scatterplot*) shows a definite pattern. This type of graph is familiar to all high-school science students who have carried out experiments that produce a set of points through which a straight line could be drawn (although radical adjustments to the data set were sometimes required for this to happen). The fact that the data falls into a line indicates that the two quantities are correlated. In doing science carefully, the mathematical tool needed to determine how good the line (or curve) fits the data set is called a *correlation coefficient*.

The data set and scatterplot in the following figure might have been produced in a simple experiment which resulted in six data points. Mathematics provides a method for finding the *line of best fit*. It is called the *method of least squares*. The details of this method must be omitted here, but most spreadsheet software and math software and some sophisticated calculators now have built-in routines that handle the tedious arithmetic involved. What results is a line or curve called a *regression line* or *regression curve*, which summarizes the correlation between two quantities. What also results is a correlation coefficient, which says how reliable the correlation is.

The correlation coefficient is a number between -1 and 1. When it equals 0, it indicates that there is absolutely no correlation between the data. When the coefficient equals -1 or 1, it means there is a perfect correlation—the points on the data set are all on the same line. A value closer to 1 or -1 obviously indicates a stronger correlation, whereas those between $-.5$ and $.5$ indicate a weak correlation.

The *coefficient of determination* is the square of the correlation coefficient. It indicates how much of the change in the second factor is explained by changes in the first. In the figure below, the line of best fit was machine-calculated. Its equation is $y = 2.4x + .933$. The correlation coefficient is 0.93487. The coefficient of determination is $(0.93487)^2 = 0.874$. Thus the fit is very good, to the extent that 87.4% of the changes in factor B are explained by factor A.

x value:	1	2	3	4	5	6
y value:	2	8	7	12	11	16

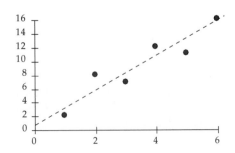

The Misuse of Statistics

Any almanac contains page after page of statistical tables, as does the annual *Statistical Abstract of the United States*. Such publications are pored over by newspaper editors, talk-show hosts, lobbyists, and political aides who try to use statistics to bolster an argument or sway public opinion. As a result, statistics is perhaps the most misused of all the sciences. When cleverly crafted bar graphs, pie charts, bell curves, and statistical averages are employed to demonstrate a trend where none exists, or to grossly overstate the extent of a social problem—whether crime, poverty, or reports of missing children—the public's attention can be diverted from more serious issues and solvable problems.

When statistics are reported in the media, it is wise to be skeptical. In considering the results of a poll, it is essential to know how many people were polled and whether the sample could have been biased. In interpreting the results of a scientific study, it is worth asking who conducted the study, how the measurements were made, and whether the results might be attributable to other factors.

Both of the following tables contain useful statistics, although they can easily be misinterpreted. Table 5.18 gives current life expectancies over a range of ages for black and white males and females. It shows that life expectancy differs between age groups, between racial groups, and between males and females. The table predicts longevity on the basis of past data. It also shows that probabilities can change as events unfold. The life expectancy of a newborn white female baby is close to 80 years, yet a white female who lives to age 80 can expect to live another 9 years. This is because those who formed the left side of the statistical bell curve (to the left of the original mean of 80) have died. But the 80-year-old's new life expectancy represents only an average; the table states nothing more than that the entire group of 80-year-old white women will average 9 more years.

Table 5.19 lists causes of death. Such tables are among the most frequently misinterpreted types of data sets. This table does provide a useful breakdown, but it should not be used to draw conclusions about individual cases. It does not say that a particular person chosen at random has a 23.5% chance of dying of cancer; it simply says that, of all deaths recorded in the Untied States (over an unspecified period), 23.5% resulted from cancer. Those deaths disproportionately occurred within high-risk groups, so the frequency should not automatically be converted to a probability. A more detailed analysis, taking into account lifestyle and hereditary factors, would give a more useful idea of an individual's risk of death from cancer.

Table 5.18. Life Expectancy

		Remaining life expectancy			
	All	*White*		*Black*	
Age in 1990	*groups*	*Male*	*Female*	*Male*	*Female*
at birth	75.4	72.7	79.4	64.5	73.6
1	75.1	72.3	78.9	64.8	73.8
2	74.1	71.4	78.0	63.9	72.9
3	73.1	70.4	77.0	62.9	71.9
4	72.1	69.4	76.0	62.0	71.0
5	71.1	68.5	75.0	61.0	70.0
10	66.3	63.5	70.1	56.1	65.1
15	61.3	58.6	65.2	51.3	60.2
20	56.5	54.0	60.3	46.7	55.3
25	51.9	49.3	55.4	42.4	50.6
30	47.2	44.7	50.6	38.2	45.9
35	42.6	40.1	45.8	34.1	41.3
40	38.0	35.6	41.0	30.1	36.8
45	33.4	31.1	36.2	26.2	32.4
50	29.0	26.7	31.6	22.5	28.2
55	24.8	22.5	27.2	19.0	24.2
60	20.8	18.7	23.0	15.9	20.5
65	17.9	15.2	19.1	13.3	17.2
70	13.9	12.1	15.4	10.7	14.1
75	10.9	9.4	12.0	8.6	11.2
80	8.3	7.1	9.0	6.7	8.6
85 and over	6.1	5.2	6.4	5.0	6.3

[Source: *Statistical Abstract of the United States*, 1995]

Any reader interested in exploring the misuse of statistics would do well to consult Darrell Huff's *How to Lie with Statistics* (W. W. Norton, 1993), a classic written in plain language.

Baseball Statistics

No fans are more obsessed with numbers than baseball fans, and this obsession has produced encyclopedic compendiums covering every imaginable baseball stat. Although television commentators often delve into the most obscure of these statistical facts ("he doesn't hit well from the left side in the late innings of afternoon games"), there are a few essential stats that any fan of the game should know. These so-called "averages," which measure the effectiveness of a batter, pitcher, or fielder, are nothing more than relative frequencies and ratios.

Table 5.19 Causes of Death in the United States

Natural Causes

Heart disease	34.0%
Cancer	23.5%
Cerebrovascular disease	6.7%
Influenza and pneumonia	3.7%
Diabetes	2.2%
Arterial diseases	2.0%
Chronic liver disease, cirrhosis	1.2%
Other diseases	19.7%
Total	93.0%

External Causes

Motor vehicle accidents	2.2%
Other accidents	2.2%
Suicide	1.4%
Homicide	1.2%
Total	7.0%

[Source: *Life Insurance Fact Book,* American Council of Life Insurance]

1. An *at-bat* is a turn at the plate that results in a hit or an out. It does not include walks, hit batters, bases awarded because of obstruction or interference, or sacrifices to advance a base runner.

2. A player's *batting average* is the ratio of hits to at-bats. It is expressed as a decimal number between 0 and 1 carried to three decimal places. In speaking of batting averages, the decimal point is ignored; thus a "300 hitter" is someone whose batting average is .300.

$$\text{batting average} = \frac{\text{no. of hits}}{\text{no. of at-bats}}$$

◆ In the 1941 season, Ted Williams of the Boston Red Sox led the American League in hitting with a batting average of exactly .400 (179 hits in 448 at-bats, which in baseball gets rounded to the third decimal place). Given the option to sit out the last day of the season to preserve a mark that had not been equaled in several decades, Williams did not hesitate to risk his average in a doubleheader against the Philadelphia Athletics. In the two games he got six hits in eight at-bats, and he finished the season with a .406 average (185 hits in 456 at-bats).

3. An *on-base percentage* is the number of times a batter reaches base divided by the number of plate appearances. Unlike at-bats, plate appearances include walks, sacrifice flies, dropped third strikes, and occasions when the batter is hit by a pitch.

4. *Slugging percentage* is a measure of a batter's ability to hit for extra bases. It is obtained by dividing the total number of bases reached safely on hits by the number of official at-bats, and it takes the form of a decimal number between 0 and 4 (though it has never risen above 1) carried to three decimal places.

• The record for slugging average is held by Babe Ruth, who slugged .847 for the New York Yankees in the 1920 season. Here is how he did it:

At-bats	Singles	Doubles	Triples	Home runs
458	73	36	9	54

Total bases = 73 + (36 × 2) + (9 × 3) + (54 × 4) = 388

$$\frac{\text{Total bases}}{\text{At-bats}} = \frac{388}{458} = .847$$

5. An *earned run average* (ERA) is a measure of a pitcher's effectiveness. An *earned run* is a run that scores as a result of an offensive or defensive action where no error is involved. An earned run average is the average number of earned runs scored against a pitcher every nine innings. It is calculated by the following formula:

$$\text{earned run average} = \frac{\text{total earned runs}}{\text{no. of innings pitched}} \times 9$$

If a pitcher does not complete a given inning, the fraction of the inning (⅓ or ⅔, depending on the number of outs) is used.

• The lowest ERA recorded in the modern era is that of Bob Gibson of the St. Louis Cardinals in 1968: 1.12.

6. A pitcher's *on-base percentage* is another measure of his effectiveness. It is the ratio of the number of hits and walks given up, divided by the number of outs plus hits and walks. It does not include intentional walks.

7. A fielder's competence is measured by his *fielding average*. This is the ratio of putouts and assists to the total number of fielding chances. It is expressed as a decimal number between 0 and 1 carried to three decimal places.

6   Business and Personal Finance

The mathematics of business and personal finance is based on the time value of money—on the difference between the face value of money and its buying power in the past or in the future. The face value of a $10 bill, for example, does not change. But if it is not invested, the amount it can buy diminishes over time; if it is invested wisely, its buying power will increase over time.

Moreover, a $10 bill can play a small part in a wide variety of activities: it can be earned, paid out, spent, gambled, invested, inherited, borrowed against, loaned out, taxed, stashed away, or given away. In each of these transactions, a value can be assigned to it, and that value will change over time.

This chapter shows how to account for a sum of money over time. It describes the mechanisms behind routine procedures that people conduct each day—paying bills, buying stocks, charging purchases, depositing checks, and paying taxes. Many of these procedures involve complicated formulas best carried out by machine. The result is a sometimes discouraging abundance of fine print on monthly statements, contracts, and application forms. In a financial world that grows more complicated every day, this is unavoidable.

Short of plunging into the mathematical intricacies of investing and borrowing, the consumer can still seek an advantage in learning some generalities. That is the purpose of this chapter. Armed with the information contained here, anyone can approach his or her banker, accountant, stockbroker, or investment analyst and ask them to clearly explain how they derive their figures. Chances are they don't know. As in the world of gambling, there is a great deal of obfuscation that goes with every type of monetary transaction. But each transaction can be clearly explained, and should be.

The fundamental concept behind all the investment decisions discussed here is interest. Once the idea of interest is understood, everything else follows.

Interest

Interest earned on investments or paid on loans can be thought of as rent charged or paid for the use of money over a period of time. Someone who opens a savings account is lending his or her money to the bank in return for compensation in the form of interest, which is calculated as a percentage of the total amount. Someone borrowing money from any type of lender is expected to pay a fee for it.

In this section, types of interest payments and ways to calculate them are examined in the context of many types of investments.

INTEREST RATES

1. *Interest* is a fee paid for the use of money. The amount loaned (or borrowed) is referred to as the *principal amount,* or simply the *principal.* The interest on the principal can be expressed either as a percent or as the actual dollar amount of fees paid. Depending on the type of loan or investment involved, a percent interest rate might be referred to as a *finance charge, discount rate, current yield,* or *annual percentage rate.*

2. As a percent, the rate of interest on an investment or a loan must refer to a time interval. Most interest rates are based on a full year; thus interest can be thought of as a yearly payment. This is usually specified as an *annual interest rate,* which is a ratio of the amount of interest paid (or earned) in one year to the principal amount invested (or borrowed) over that time.

$$\text{annual interest rate} = \frac{\text{interest earned in one year}}{\text{amount invested}}$$

- When $100 invested for one year earns $8 in interest, the annual interest rate is $8/100$, or 8% per year.

- If a bank lends $500 for one year and charges $45 in interest, the annual interest rate is ⁴⁵⁄₅₀₀, or ⁹⁄₁₀₀, or 9%.

3. On installment loans, a monthly statement will often refer to a *daily interest rate,* the interest charge for one day's use of money. This is the annual rate divided by 365. The *monthly rate* is the annual rate divided by 12, the *weekly rate* is the annual rate divided by 52, and so on.

Φ ———————————————————————— Φ

CALENDAR YEAR VS. FISCAL YEAR

A true solar year contains about 365¼ days, while the calendar year contains 365 (except in leap years). These discrepancies may seem relatively unimportant, but the maxim that "time is money" is literally true in the case of compound interest. One day's interest on $1,000,000 calculated at 8% per year is about $220. This might not seem like much. Even if one day's interest were forfeited each year over 30 years, the total interest lost would be only $220 × 30, or $3600. But this doesn't take into account the time value of money. If a $200 deposit were made each year into an investment account paying 8% per year, the principal would be worth almost $23,000 after 30 years. In other words, one day can make a big difference in the amount of money that will ultimately change hands.

Traditionally, commercial banks have calculated interest based on a 360-day year (consisting of twelve 30-day months). Although this practice is still observed in some bond trading houses, most investments earn interest over a 365-day year. U.S. Treasury bills are a holdover; unlike Treasury bonds or notes, they earn interest over a 360-day year (see page 300 to see how interest on Treasury notes and bills is reconciled). To the average investor, this should not be much of an issue, since the dollar amounts at stake are small; but for large investors, the number of days in a "year" is worth investigating.

Φ ———————————————————————— Φ

DISCOUNT RATE, T-BILL RATE, AND PRIME LENDING RATES

1. The *discount rate* is the rate at which commercial banks may borrow short-term funds from the Federal Reserve Bank (the nation's central bank)—that is, the interest rate that commercial banks have to pay for money. Set by the Federal Reserve Board, a presidentially appointed board of directors who govern the country's monetary policies, it is a key economic indicator, one of the principal determinants of the prime lending rates set by the country's leading banks.

2. The *prime lending rate* (also known simply as the *prime*) is the interest rate banks use when lending to their most creditworthy customers. Although each bank may set its own prime rate, in practice most commercial banks follow the rates set by the major banks. The prime rate published daily in the *Wall Street Journal* is an average based on the rates of corporate loans issued by 30 of the nation's largest banks. In addition, prime lending rates are calculated for individual countries. The *London Interbank Offered Rate (LIBOR)*, which is based on five European banks, is a kind of international prime rate.

Banks set their mortgage, equity loan, and personal loan rates using simple formulas based on the prime rate. Equity loans, for example, are often set at 2 points above the prime, credit-card rates anywhere from 6 to 10 points above the prime. (In general, a *point* refers to a percentage point, or to 1% of some principal amount.)

3. There are several other interest rates that act as key economic indicators by setting the cost of money. The *federal funds rate* is an interest rate for cash reserves held at Federal Reserve district banks. When one bank has excess reserves, it may lend funds to a bank that has a deficit. Although the rate they can charge is determined by the market, it is influenced by the Federal Reserve Board, which sets the minimum of reserves that banks must maintain. The federal funds rate is considered the most sensitive indicator of the direction that bank interest rates will go. Changes in the discount rate and the prime rate usually follow the federal funds rate.

4. The *T-bill rate* is the interest rate that the government pays on short-term debt obligations known as *Treasury bills*. There are several types of T-bills, which can be distinguished by their terms of maturity—three months (13 weeks), six months (26 weeks), and one year (52 weeks). Interest rates for each type of T-bill are set every Monday. They are calculated from the discount price at which each type of bill is sold at auction.

Bank rates may also be keyed to the T-bill rate. For example, many banks set their variable mortgage rates a few percentage points higher than the T-bill rate. Rates of short-term certificates of deposit offered by commercial banks are also based on the T-bill rate.

5. The Fed's discount rate is believed to act as a check on inflation. When the Fed raises interest rates, money becomes more expensive to borrow and the money supply is said to "tighten." This has the effect of reducing spending and keeping prices in check. A low prime rate makes money more easily available and tends to encourage spending. According to economists, this promotes steady economic growth. The Fed has several means with which to regulate the nation's economy, but this kind of intervention is not an exact science.

Calculating Interest Payments

Banks are required by law to post their interest rates on all investments and loans. In such listings, the first figure given is the *annual interest rate*. The second is the *effective rate* or *effective yield*, which is higher. This discrepancy exists because interest, while stated as an annual payment, is often paid (or collected) several times in a single year. As a result, interest that has been added to the principal begins to earn interest, which *compounds* the total interest payment. There are several ways of calculating interest. They are:

> *simple interest:* a one-time fee paid at the end of a contract
> *discount interest:* a one-time fee paid at the beginning of a contract
> *compound interest:* a fee calculated as a series of payments made at regular intervals during a contract

Interest payments are almost always calculated today by means of computers, business calculators, or interest tables. These employ the mathematical formulas given below.

1. Simple Interest *Simple interest* refers to a one-time calculation of a percentage of a principal. It is paid at the end of a time period specified by a lending agreement or contract. In a simple scenario, a person who borrows $100 for one year with simple interest of 6% per year would have to repay the $100 principal plus an interest charge amounting to 6% of $100, which is $6.

All simple interest is calculated using the formula

$$I = P \times r \times t$$

where P is the principal amount, r is the annual interest rate (expressed as a decimal), and t is the time in years. Although few investments or loans are calculated using simple interest, certain types of bonds do pay simple interest annually or semiannually on the face value of the bond.

> ◆ A 12½% bond has a face value of $1000. If simple interest is paid as a dividend each year, the annual payment is calculated as 12½% of $1000, or $125. After 5 years the total interest earned would be
> $$I = \$\,1000 \times 0.125 \times 5 = \$625$$

2. Discount Interest A *discount* refers to an interest charge that is collected at the beginning of the term of a loan. The amount of the discount is typically subtracted from the loan amount. A $5000 student loan with a 10% discount, for example, puts $4500 in the student's hands at the outset of the contract. The $500 discount (10% of $5000) is the interest charge. When the student finishes school, the $5000 becomes due in full after six months. If not paid in full at that time, the debt converts to an installment debt, which entails additional interest charges paid at regular intervals over a given number of years (usually five or ten).

Another type of discounted debt is a U.S. Treasury bill. These are issued in denominations starting at $10,000. All T-bills are sold at a discount, which means that the purchase price is lower than the face value. They may be redeemed for the face value when they have matured. Three- and six-month T-bills are sold at auction every Monday, and one-year bills are auctioned once a month. T-bills may also be sold on secondary markets without a great loss of value through commissions. The difference between what brokers charge for T-bills and what they will pay for them is usually in the range of 2 to 4 *basis points* (that is, 2 to 4 *hundredths* of a percent of the face value).

A three-month Treasury bill worth $10,000 upon redemption might sell for $9800. The discount, or interest earned by the investor, is $200. For a more detailed discussion of Treasury bills, see page 300.

3. Compound Interest When the interest earned on an investment is paid out at regular intervals (rather than as a single payment at the beginning or end of a contract), it can be added onto the principal amount. When this is done, all subsequent interest payments are based on the principal and the accumulated interest. That is, the accumulated interest earns interest. This is called *compounding*.

For example, assume an investment of $100 earns 10% per year compounded yearly. Table 6.1 shows how the investment grows if the interest earnings are automatically added onto the principal.

Table 6.1. Compound Interest on $100

Year	Starting balance	Interest earned	New balance
1	100	$100 \times 0.10 = 10$	110.00
2	110	$110 \times 0.10 = 11$	121.00
3	121	$121 \times 0.10 = 12.1$	133.10
4	133.10	$133.10 \times 0.10 = 13.31$	146.41
5	146.41	$146.41 \times 0.10 = 14.64$	161.05

After five years the interest amounts to $61.05. If simple interest had been paid instead of compound interest, the interest payments would have totaled $50 (because $I = P \times r \times t = \$100 \times 0.10 \times 5 = \$50$). The interest rate is the same in both cases, but compounding brings in the extra $11.05 as interest earned on accumulated interest.

While the compounding of interest works to the advantage of the investor (since compound interest amounts to more than simple interest), it works to the disadvantage of the borrower. Compounding increases the cost of borrowing. Just as interest earned on a savings account is rolled into the principal, resulting in interest earning interest, the amount of interest owed on a debt is also rolled into the principal, so that the borrower pays interest on the interest owed.

Whether money is borrowed or invested, the interest earned (or paid) on it may be compounded yearly, semiannually (every six months), quarterly (four times a year, or every three months), monthly, biweekly, weekly, or daily. In theory, interest can be compounded any number of times at regular intervals during a year. The number of compoundings determines the total amount of interest earned; the more frequently interest is compounded, the faster the investment grows.

CALCULATING COMPOUND INTEREST

1. Compounding means adding an interest payment onto a principal amount. Mathematically, this is the same as calculating a *markup* (see page 15). Assume a percent increase is given in decimal form as r (to find r, divide the percent interest rate by 100). To increase or *mark up* an amount by a rate r, multiply the amount by $(1 + r)$.

♦ If a salary of $20,000 is increased by 5%, the new salary is $20,000 × (1 + 0.05), or $21,000.

If interest is automatically added onto the principal amount, it acts like a markup.

♦ A 5% interest payment on $20,000, added directly to the principal, results in a new principal amount of $20,000 × (1 + 0.05), or $21,000.

2. Just as the calculation of an initial interest payment involves multiplying the principal by $(1 + r)$, each subsequent compounding of interest can be calculated by multiplying again by the same factor.

♦ The second compounding of the principal of $20,000 is calculated by multiplying the result of the first compounding ($21,000) by (1 + 0.05). In this calculation, $21,000 is written as $20,000 ×

(1 + 0.05) in order to show a pattern. The principal after the second compounding would be

$$[20,000 \times (1 + 0.05)] \times (1 + 0.05)$$

or

$$20,000 \times (1 + 0.05)^2$$

After three compoundings, the principal amount would be

$$20,000 \times (1 + 0.05)^3$$

After n compoundings, the principal would be

$$20,000 \times (1 + 0.05)^n$$

3. The process of compounding involves multiplying a principal amount by an *interest accumulation factor* of the form $(1 + i)$, where i is the interest rate (in decimal form) paid *for that compounding period.*

If interest is compounded annually, and r is the annual interest rate, then the interest accumulation factor is $(1 + r)$. If interest is compounded monthly, then the interest rate used in the interest accumulation factor must be converted to a monthly rate. The monthly rate is the annual rate divided by 12. Thus $(1 + i)$ would take the form $(1 + r/12)$.

4. In general, if the annual interest rate is r and interest is compounded m times per year, then the interest accumulation factor used for each compounding is $(1 + r/m)$. In compounding the interest on an investment of P dollars, multiply P by the interest accumulation factor *one time for each compounding.* This leads to a general formula for accumulated compound interest known as the *future value formula.*

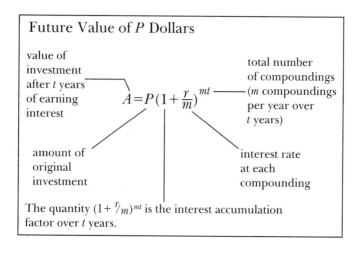

Future Value of P Dollars

value of investment after t years of earning interest

total number of compoundings (m compoundings per year over t years)

$$A = P\left(1 + \frac{r}{m}\right)^{mt}$$

amount of original investment

interest rate at each compounding

The quantity $(1 + r/m)^{mt}$ is the interest accumulation factor over t years.

• Assume $750 is invested at 8% per year for five years with monthly compounding. Then $P = 750$, $r = .08$, $m = 12$, and $t = 5$. At the end of five years, the investment would be worth

$$A = 750\left(1 + \frac{.08}{12}\right)^{12 \times 5} = \$1117.38$$

This calculation can be done on a scientific calculator using the key marked x^y. First compute the value of $(1 + .08 \div 12)$. Then press $\boxed{x^y}$ followed by 60 (which is 12×5). The result (about 1.4898) is the total interest accumulation factor over five years of compounding. Multiply by $750 to get the final answer of $1117.38.

FUTURE VALUE AND PRESENT VALUE

Once invested, a sum of money will increase in value over time (assuming it is invested wisely). If left under a mattress, it retains its face value but diminishes in buying power due to the effects of inflation. Consequently, $500 in cash today is worth more than a promise to pay $500 five years from now. Assuming an annual interest rate of 5% and monthly compounding, $500 invested *now* would turn out to be worth

$$500 \times \left(1 + \frac{.05}{12}\right)^{12 \times 5} = \$641.68$$

five years from now.

In this calculation, $500 is referred to as the *present value* of the money, and $641.68 is the *future value* of the money. The formula that relates the present and future value of money is the compound interest formula

$$A = P \times \left(1 + \frac{r}{m}\right)^{m \times t}$$

where P represents the present value and A represents the future value. The formula above, when solved for the present value P, looks like this:

$$P = A \div \left(1 + \frac{r}{m}\right)^{m \times t}$$

In comparing the formulas, notice that a division sign has replaced the multiplication sign and that the positions of A and P have been reversed. Together, these formulas can be used to solve two types of compound interest problems.

The future value problem: How much is P dollars worth in t years if it earns an interest rate r and is compounded m times per year?

The present value problem: How much money should be invested *today* if it is to be worth A dollars t years from now?

The first type of problem has been dealt with. An example of the second is given below.

• In order to have $1000 in a bank account eight years from now, the amount that should be invested *now* at 6% per year with weekly compounding would be

$$1000 \div \left(1 + \frac{.06}{52}\right)^{52 \times 8} = \$618.95$$

To carry this out on a scientific calculator, first calculate the value of (.06 ÷ 52) and add 1. Then press $\boxed{x^y}$ followed by 416 (which is 52 × 8). To divide 1000 by this number, press the reciprocal key (marked $\boxed{1/x}$ or $\boxed{x^{-1}}$) and multiply the result by 1000. The result given above should appear.

When a loan is *discounted* using compound interest, the present value of the loan amount can be calculated using the present value formula, but it may also be discounted using simple discount interest.

• If a $10,000 loan is discounted at 6% and comes due in six months (0.5 years), the borrower would receive the present value of $10,000, which is

$$P = 10,000 \div \left(1 + \frac{.06}{2}\right)^{2 \times 0.5} = \$9708.74.$$

If the same note were discounted using simple interest, the discount would be (10,000) × (0.06) × (0.5) = $300.

In the example above, the first method of discounting is called a *true discount,* the second a *bank discount.* The first is more favorable to the borrower because it puts more money in his or her hands and thus entails a smaller interest payment. The second method, however, is the more standard practice.

EFFECTIVE YIELD AND ANNUAL PERCENTAGE RATES (APR)

Table 6.2 compares the interest earned on $100 invested for one year at 10% per year where different compounding schemes are followed. The formula used to calculate future value is

$$A = P \times \left(1 + \frac{r}{m}\right)^{m \times t}$$

With $P = \$100$, $r = 10\% = 0.1$, and $t = 1$ year, the formula simplifies to

$$A = 100 \times \left(1 + \frac{0.1}{m}\right)^m$$

This formula, with different values for m, generates the figures shown below.

Table 6.2. Effective Yield for Various Compounding Periods

No. of compoundings (m)	Value of $100 after 1 year (A)	Interest earned	Effective interest rate
1	110	10	10%
2	110.25	10.25	10.25%
4	110.38	10.38	10.38%
12	110.47	10.47	10.47%
52	110.51	10.51	10.51%
365	110.52	10.52	10.52%
infinite	110.52	10.52	10.52%

In this table there are two patterns worth noting. First, the amount of interest earned increases with the frequency of compounding. Thus the stated annual interest rate is increased through the process of compounding into a higher interest rate. This higher rate, calculated over one year, is called the *effective annual interest rate,* or simply the *effective yield.*

Secondly, the effective yield does not increase indefinitely as the number of compoundings increases. It has an upper limit. In this example, the limit is precisely $(e^{0.1} - 1)$, where $e^{0.1}$ is calculated on a calculator using the key marked e^x. (The value of $e^{0.1} - 1$ works out to 0.105171, which is rounded in the table to 10.52%.)

1. The *effective yield* of any investment depends on the stated annual interest rate and the number of compoundings per year. It represents the actual amount of interest earned (or paid) in one year when the effect of compounding is taken into account. The formula for effective yield is:

$$\text{effective yield} = \left(1 + \frac{r}{m}\right)^m - 1$$

The greatest possible yield (assuming an infinite number of compoundings per year) is $e^r - 1$.

- A passbook savings account earning 5% per year with daily compounding has an effective yield of

$$\left(1 + \frac{.05}{365}\right)^{365} - 1 = 0.0512675, \text{ or about } 5\frac{1}{8}\%$$

2. While effective yield applies to interest earned on invested money, the term *APR*, or *annual percentage rate*, applies to interest owed on loans. The idea is essentially the same. When a debt is paid off in installments, interest charges accumulate. In time, interest is owed on unpaid interest. This compounding causes the effective rate of interest to be higher than the stated rate. If a loan contains no charges other than compound interest, the APR is the same as the effective yield. But most loans include other costs that technically constitute interest. These may take the form of application fees, borrower's insurance, points (a form of prepaid interest in the amount of 1% of the borrowed amount), maintenance fees, and so on. The APR is calculated by adding these charges into the cost of the loan as interest. Thus the APR is always greater than the stated annual interest rate used for compounding purposes, and it is usually greater than the effective yield. By law, every lender must disclose the APR in large bold type.

   • A variable-rate mortgage is advertised at 6.75% per year for the first five years, after which it adjusts to 2 percentage points over the T-bill rate (see page 260). Because of loan origination fees, interest compounding, and monthly private mortgage insurance, the bank lists an APR of 7.85%. Another bank advertising a rate higher than 6.75% may have a lower APR than this because it is willing to pick up some of the application fees and closing costs. Therefore it is advisable to shop for mortgages and other loans by comparing APR and not merely the listed annual interest rate.

Table 6.3. Summary of Compound Interest Formulas

   P = present value (or principal amount)

   A = future value of P

   r = annual interest rate in decimal form (divide the annual percent rate by 100)

   m = number of compoundings per year

   t = number of years

$$\text{Future Value: } A = P \times \left(1 + \frac{r}{m}\right)^{m \times t}$$

$$\text{Present Value: } P = A \div \left(1 + \frac{r}{m}\right)^{m \times t}$$

$$\text{Effective Interest Rate} = \left(1 + \frac{r}{m}\right)^{m} - 1$$

Greatest Possible Effective Yield $= e^{r} - 1$

DOUBLING TIME: THE LAW OF 70

1. Any investment that earns interest will double in size over a certain number of years, regardless of the amount invested, and it will double again over the same time interval. At an annual rate of i percent with m compoundings per year, the doubling time is given by the formula

$$\text{doubling time} = ln(2) \times \frac{1}{m} \div ln\left(1 + \frac{r}{m}\right)$$

where ln stands for the natural logarithm (see Chapter 1) and r is the decimal form of the annual interest rate i.

• Suppose $5000 is invested at 6% per year compounded monthly. How long will it take to double in value?

$$\text{doubling time} = ln(2) \times \frac{1}{12} \div ln\left(1 + \frac{.06}{12}\right)$$
$$= 11.58 \text{ years}$$

2. A much simpler doubling-time formula assumes that interest is compounded continuously (meaning at every instant of time). This formula provides a good approximation that works in any compounding situation.

$$\text{doubling time} = \frac{ln(2)}{r}$$

• In the example above, the simpler doubling-time formula gives a doubling time of $^{ln(2)}\!/_{.06}$, which equals 11.55 years. This differs from the exact answer by about 11 days.

3. The simplified doubling-time formula leads to similar formulas for tripling time, quadrupling time, and so on.

$$\text{doubling time} = \frac{ln(2)}{r}$$
$$\text{tripling time} = \frac{ln(3)}{r}$$
$$\text{quadrupling time} = \frac{ln(4)}{r}$$
$$\text{quintupling time} = \frac{ln(5)}{r}$$

Table 6.4. Doubling, Tripling, Quadrupling, and Quintupling Times of Investments

Annual interest rate	Doubling time	Tripling time	Quadrupling time	Quintupling time
4%	17.32	27.47	34.66	40.24
5%	13.86	21.97	27.73	32.18
6%	11.55	18.31	23.10	26.82
7%	9.90	15.69	19.80	22.99
8%	8.66	13.73	17.33	20.12
9%	7.70	12.21	15.40	17.88
10%	6.93	10.99	13.86	16.09
11%	6.30	9.99	12.60	14.63
12%	5.78	9.16	11.55	13.41
13%	5.33	8.45	10.66	12.38
14%	4.95	7.85	9.90	11.50
15%	4.62	7.32	9.24	10.73
16%	4.33	6.87	8.66	10.06
17%	4.08	6.46	8.15	9.47
18%	3.85	6.10	7.70	8.94
19%	3.65	5.78	7.30	8.47
20%	3.47	5.49	6.93	8.04

4. The Law of 70 A convenient rule of thumb says that the doubling time for any investment can be found by dividing the percent annual interest rate into 70.

$$\text{doubling time} = \frac{70}{\text{percent interest rate}}$$

• An investment earning 7% per year will double in about $70/7 = 10$ years. At 14%, it will double in about $70/14 = 5$ years. Comparing these to the values in the table shows that the Law of 70 is adequate for most purposes.

Annuities and Rents

1. An *annuity* is a sequence of equal payments made at regular intervals. This includes anything involving fixed weekly, monthly, or yearly payments. Salary, insurance premiums, rent, and mortgage payments are all forms of annuities. An annuity is also a form of investment. Insurance companies offer policies that not only provide life insurance but also accumulate cash value with each payment made.

Because annuity payments are spread out over time, their total future value is greater than the sum of the individual payments. Moreover, because the payments are made *periodically,* each payment accumulates interest over a time interval different from that of other payments. A stripped-down example will serve to illustrate this.

♦ If a payment of $100 is made each month into an account that earns 6% interest compounded monthly, the total amount deposited over a year will be $1200. However, the first $100 deposit will earn interest over the entire 12 months, the second $100 over 11 months, the third over 10 months, and so on. These principal and interest amounts can be calculated individually using the future value formula, which says that future value $A = P(1 + r/m)^{mt}$. The value after 1 year of the successive payments would be:

payment 1:
$$A = 100 \times \left(1 + \frac{0.06}{12}\right)^{12} = 100 \times (1.005)^{12} = \$106.17$$

payment 2:
$$A = 100 \times \left(1 + \frac{0.06}{12}\right)^{11} = 100 \times (1.005)^{11} = 105.64$$

payment 3:
$$A = 100 \times \left(1 + \frac{0.06}{12}\right)^{10} = 100 \times (1.005)^{10} = 105.11$$

$$. . .$$

payment 10:
$$A = 100 \times \left(1 + \frac{0.06}{12}\right)^{3} = 100 \times (1.005)^{3} = 101.51$$

payment 11:
$$A = 100 \times \left(1 + \frac{0.06}{12}\right)^{2} = 100 \times (1.005)^{2} = 101.00$$

payment 12:
$$A = 100 \times \left(1 + \frac{0.06}{12}\right)^{1} = 100 \times (1.005)^{1} = 100.50$$

The total value of the year's worth of payments is:

$$S = 100 \times (1.005)^1 + 100 \times (1.005)^2 + 100 \times (1.005)^3$$
$$+ . . . + 100 \times (1.005)^{11} + 100 \times (1.005)^{12}$$
$$= \$1239.72$$

Φ ———————————————————————————— Φ

THE HALF-LIFE OF A DOLLAR

The expression "good old days" conjures up a time when a good cigar cost a nickel, a good car cost $700, and a good house could be had for a few thousand dollars. Of course, those days and those dollars are long gone. What we have now is sometimes referred to as the "fifty-cent dollar" or the "two-bit dollar." These figures—50 cents and 25 cents— represent one and two half-lives of a dollar, respectively. This raises the question: How long does it take for the dollar to lose half its value?

The half-life of a dollar depends on the rate of inflation. If inflation is constant, then the half-life of money is the same as the doubling time of prices. When prices double, the buying power of a dollar is cut in half. Thus, by the Law of 70,

$$\text{half-life of a dollar} = \frac{70}{\text{annual inflation rate}}$$

If the average annual rate of inflation were 5%, then the half-life of money would be about $70/5$, or 14 years. Thus in 28 years we would find ourselves stuck with a two-bit (25-cent) dollar.

Φ ———————————————————————————— Φ

Fortunately, the calculation of the future value of an annuity can be summarized in a single formula. If we assume a regular payment of P dollars is made n times per year into an account in which interest is compounded at the end of each payment period, the total future value of these n payments would be given by the formula

$$FV = P \times [(1 + i)^n - 1] \div i$$

where

$FV = $ total future value of the annuity
$P = $ amount of periodic payment
$n = mt = $ total no. of payments (no. of compoundings times no. of years)
$i = i/m = $ interest rate per compounding period

Note that for this formula to work, m, the number of *compoundings* per year, must equal the number of *payments* made per year. (To use a different compounding period requires a more complex formula.) In practice, the task of calculating annuity sums and payments is entrusted to computers or annuity tables.

• If a father deposits $100 for his son each month into an account earning 5% per year, compounded monthly, what will the account be worth after 18 years?

$$\text{future value} = 100 \times \left[\left(1 + \frac{.05}{12}\right)^{12 \times 8} - 1 \right] \div \left(\frac{.05}{12}\right) = \$34,920.$$

2. Sinking Funds Instead of calculating the future value of a series of regular payments, it may be more useful to calculate the installment payment that must be made if an annuity is to grow to a certain amount over a fixed interval of time. This situation—meeting an investment goal through a series of regular payments—is called a *sinking fund.*

If the formula for the future value of an annuity is rearranged, it can be used to solve for the installment payment in a sinking fund. The result is

$$P = FV \times i \div [(1 + i)^n - 1]$$

where P is the installment, FV is the future value of the annuity, i is the interest rate per compounding period (which is the same as the installment period), and n is the total number of installment payments.

• A mother who wishes to set aside $50,000 for her daughter's college tuition sets up a monthly payment plan over 15 years. Assuming a rate of growth of 8% per year (using monthly compounding), her monthly payment should be

$$50,000 \times \left(\frac{.08}{12}\right) \div \left[\left(1 + \frac{.08}{12}\right)^{12 \times 15} - 1 \right] = \$144.49$$

3. Present Value of an Annuity Because the promise to pay a sum of money sometime in the future is worth less than the same amount of cash right now, a promise to pay a fixed sum over a series of regular intervals is worth less than the total value of the payments. A promise to pay $100 a month over the next year, for example, is worth less than $1200 right now. How much less is determined by the present value of each of the $100 payments. The first payment needs to be discounted over a month, the second over two months, the third over three months, and so on.

• Assume that $100 is to be paid on the last day of the month over an entire year. The present value of each payment, assuming 6% interest compounded monthly, is:

payment 1:
$$P = 100 \times \left(1 + \frac{0.06}{12}\right)^{-1} = 100 \times (1.005)^{-1} = \$99.50$$

payment 2:
$$P = 100 \times \left(1 + \frac{0.06}{12}\right)^{-2} = 100 \times (1.005)^{-2} = 99.01$$

$$\bullet \quad \bullet \quad \bullet$$

payment 11:
$$P = 100 \times \left(1 + \frac{0.06}{12}\right)^{-11} = 100 \times (1.005)^{-11} = 94.66$$

payment 12:
$$P = 100 \times \left(1 + \frac{0.06}{12}\right)^{-12} = 100 \times (1.005)^{-12} = 94.19$$

The sum of these 12 present values would be:
$$S = 100 \times (1.005)^{-1} + 100 \times (1.005)^{-2} + \ldots$$
$$+ 100 \times (1.005)^{-11} + 100 \times (1.005)^{-12}$$
$$= \$1161.89$$

As in the case of the future value of an annuity, a single formula will generate such a sum.

$$PV = \text{payment} \times \left[1 - \frac{1}{(1 + i)^n}\right] \div i$$

where

PV = the present value of the annuity

n = the number of payments

$i = \dfrac{r}{m}$(annual interest rate divided by no. of compoundings) = the interest rate per compounding period

Using this formula, the example given above is calculated in the following way:

$$\text{present value} = \$100 \times \left[1 - \frac{1}{\left(1 + \frac{.06}{12}\right)^{12}}\right] \div \left(\frac{.06}{12}\right)$$

$$= \$100 \times \left[1 - \frac{1}{1.005^{12}}\right] \div (.005)$$

$$= \$1161.89$$

In this formula, as in the future value formula given earlier, the compounding period must be the same as the payment period. If it is not, this formula will at least provide a good estimate.

4. Amortization of Debt When a loan amount is paid off through a series of regular installments, it is said to be *amortized,* which literally means "killed off." Each payment retires some of the unpaid balance, and the rest constitutes interest on the balance. This is what happens when a mortgage payment is made. The principal owed is reduced with each payment. This contributes to the amount of equity the owner has in the property (see page 277).

This is also the premise of any time payment plan, or installment plan. A borrower is advanced a sum of money or credited with a purchase, and then makes periodic payments in order to retire the debt. The periodic payments consist of principal and interest. That is, with each payment a part of the debt is retired and an interest charge is paid. At any time the principal may be paid off in full (although there may be a prepayment penalty). Loans of this type are the most common loans. They include mortgages and credit-card debt installments. By order of the Truth in Lending Act of 1968, all installment loans are required to disclose their annual percentage rate (APR).

5. Size of Periodic Payments for an Installment Loan The size of an installment payment on any loan (or purchase) is based on the amount borrowed (the principal), the stated interest rate (r), and the total number of installment payments to be made (n). In the following formula, i designates the interest rate *per compounding period* stated as a decimal. It is equal to the annual interest rate r divided by the number of compoundings per year. (The number of compoundings is often the same as the number of payments per year, but not necessarily. Mortgages, for example, are usually compounded twice a year even though the payments are monthly.) The value n is the number of payments made over the life of the loan. If the payments are monthly, for example, multiply the term of the loan in years by 12 to get n. Here is the formula:

$$\text{payment size} = (\text{principal}) \times i \div \left[1 - \frac{1}{(1 + i)^n} \right]$$

Thus, to find the size of your installment payment:

i. Convert the annual interest rate to a decimal, then divide by the number of payments per year. (This gives i, the rate per installment period.)

ii. Add 1, then raise the result to the power n, which is equal to the number of payments over the life of the loan. (Use the $\boxed{x^y}$ key on your calculator.)

iii. Find the reciprocal of this result (use the $\boxed{1/x}$ key), and subtract it from 1.

iv. Take the reciprocal again (using the $\boxed{1/x}$ key).

v. Multiply the result by the principal, and then by the value of i as computed in step 1. The result is your monthly payment.

◆ A family purchases a house for $120,000 with a 20% down payment. The size of their mortgage is $120,000 − (0.20) × ($120,000) = $120,000 − 24,000 = $96,000. Assuming they settle on a 30-year mortgage (which would entail a total of 360 monthly payments) with an annual rate of 9% (and thus a monthly rate of $.09/12$, or .0075), their monthly mortgage payment would be

$$(96,000) \times (.0075) \div \left[1 - \frac{1}{(1.0075)^{360}} \right] = \$772.44$$

The figures in Table 6.5 are calculated from this formula.

Mortgage tables provide the easiest way to calculate monthly payments. Any bank can provide a mortgage table that lists monthly payments for a wide variety of mortgaged amounts, time periods, and interest rates.

6. Amortization Schedules An *amortization schedule* shows how a debt is retired, payment by payment. It shows how much of a payment constitutes interest and how much goes toward retiring the debt. By breaking each payment into interest and principal components, it shows the total amount of interest paid at any point during the repayment of the debt. Table 6.5 consists of an amortization schedule for a $1000 loan paid back in six monthly payments. The interest rate is 1% per month, which is equivalent to 12% per year. The monthly payment is calculated using the formula given above, where the rate per month (designated i) is 1% (or 0.01), and the total number of payments (designated n), is 6.

$$\text{payment size} = (1000) \times (0.01) \div \left[1 - \frac{1}{(1.01)^6} \right] = \$172.55$$

At the end of the first month, a payment of $172.55 is due. This includes interest in the amount of 1% of $1000, which amounts to $10. Thus the first payment consists of $10 in interest and $162.55 of debt reduction, which reduces the debt to $837.45.

At the end of the second month, the interest amounts to 1% of this new balance, which works out to $8.37. The interest payment will always be 1% of the previous unpaid balance. With each payment the unpaid balance diminishes, as does the size of the interest payment.

The last payment usually has to be adjusted by a small amount in order to retire the unpaid balance. This is because each computation involves some roundoff error.

Table 6.5. An Amortization Schedule

Payment number	(Payment	= Interest	+ Principal)	Unpaid balance
1	$172.55	$10.00	$162.55	$ 837.45
2	$172.55	8.37	164.18	673.27
3	$172.55	6.73	165.82	507.45
4	$172.55	5.08	167.47	339.98
5	$172.55	3.40	169.15	170.83
6	$172.55	1.70	170.83	
Totals	$1035.30	$35.28	$1000.00	

7. Equity *Equity* is the term used to describe the cash value of the owner's share in any investment. One such investment is real estate. The owner's equity in a house is calculated by this formula:

equity = (current market value) − (unpaid mortgage balances)

When a buyer puts a 20% down payment on a $100,000 house, his initial equity is $20,000—the actual amount of cash put in. The bank supplies the other $80,000, and this, the outstanding amount of the mortgage, represents its share of ownership. As the debt is retired, the buyer acquires more and more equity, which can eventually be used as collateral on an *equity loan* (or *second mortgage*). Also, as the owner puts money into repairing or renovating the house, its market value increases, as does the owner's equity.

However, not every dollar of the monthly mortgage payment is equity. At the outset of the loan, a large part of it is interest. The federal government has turned this to the homeowner's advantage by making all of the mortgage interest tax-deductible. The amount of each payment that constitutes interest gradually declines with each mortgage payment. Therefore the deductions are greatest at the outset of the loan. At any point during the repayment of a mortgage, the amount of the mortgage payment that constitutes interest can be computed by multiplying the unpaid balance by a percent.

- If a 30-year mortgage has a fixed annual interest rate of 7.5%, and the unpaid balance is $102,000, the next mortgage payment will contain a month's worth of interest on this balance. The monthly rate is (0.075 ÷ 12), or 0.00625. The next payment will contain an interest charge in the amount of (0.00625 × 102,000) = $637.50.

This amount (as well as the interest paid in the other 11 months of the year) can be claimed as a tax deduction.

Table 6.6. Summary of Annuity Formulas

FV = total future value

PV = total present value

P = amount of periodic payment

n = total number of payments

r = annual interest rate (in decimal form)

m = number of compoundings

$i = \dfrac{r}{m}$ = interest rate per compounding period

Future value of an annuity:
$$FV = P \times [(1 + i)^n - 1] \div i$$

Sinking fund (calculating the periodic payment needed to reach an investment goal):
$$P = FV \times i \div [(1 + i)^n - 1]$$

Present value of an annuity:
$$PV = P \div i \times \left[1 - \frac{1}{(1 + i)^n} \right]$$

Amortization Formula (calculating the size of the monthly payment on a loan):
$$\text{periodic payment } P = (\text{principal}) \times i \div \left[1 - \frac{1}{(1 + i)^n} \right]$$

Mortgages

Buying a home is the single largest purchase most people will ever make. It would be beyond the means of most consumers if it were not for *mortgages*, which are installment loans for the purchase of real estate. Unlike other types of loans, mortgages carry a considerable tax advantage—all of the interest paid on a home mortgage is tax-deductible. And because real estate is a relatively dependable form of investment, banks, credit unions, and mortgage companies actively compete with each other to offer the most attractive terms.

Over the last 20 years, lending institutions have come up with a confusing array of options. Where once the 30-year fixed-rate mortgage reigned supreme, many combinations of repayment schemes are now possible. In addition, the mortgage industry has ridden a wave of refinancing over the last decade as interest rates have fallen. *Second mortgages* or *equity loans* have become increasingly popular. Because the interest on equity loans (money borrowed against the equity an owner has accumulated in a property) is tax-deductible, many homeowners have applied for equity loans instead of personal loans, which are far more expensive.

There are two basic types of mortgage, those with fixed interest rates (and therefore fixed monthly payments) and those with variable interest rates (and potentially varying monthly payments). Either type is offered over time periods ranging from five years to 30 years or more. The debt is amortized over this time, but the possible amortization schedules are numerous. Some mortgages are designed so that the payments are small at the outset. Others involve a sizable jump in payments after an initial grace period. But all mortgages are nothing more than annuities, installment loans governed by the same present and future value formulas discussed in the previous section. (The formula for calculating the amount of a monthly mortgage payment, which is also the formula for any installment loan, is given on page 275.)

INTEREST RATES AND POINTS

1. Mortgages are offered at *annual interest rates.* Rates offered by banks are competitive and tend to fluctuate in anticipation of changes in such indexes as the Federal Reserve discount rate, the average yield on Treasury securities, and the federal funds rate.

Banks will usually allow a borrower to *lock in* at the rate that is in effect when a mortgage application is filled out. (The *commitment letter,* the lender's written statement of the terms of the loan it will offer, refers to such an agreement as a *rate lock* or *interest rate commitment.*) This is particularly advantageous during periods of rising interest rates. If the rate falls during the time it takes to close on the house, some banks will automatically offer the lower rate; others will charge a fee for it.

2. Banks maintain a *margin* between the interest rates they pay on bank accounts and the interest rates they charge for mortgages. This is one way they make money. The margin is not a fixed amount; banks will offer different rates for almost every type of loan they make. Their lowest rates are usually for their shortest-term variable-rate mortgages; the highest are for personal loans.

3. It is often possible to get a lower interest rate by *buying down* the rate—that is, by paying fees referred to as *points,* a *point* being a fee that is treated (for tax purposes) as prepaid interest, equal to 1% of the amount borrowed. Points may be charged by a bank as a *loan origination fee*. They may also be paid by a buyer who wishes to buy into a lower interest rate. The seller may even be willing to pay points on behalf of the buyer if it will help the sale go through.

> ◆ A couple buys a $150,000 house with a 20% down payment. The amount borrowed is
>
> $150,000 − (.20 × 150,00) = $150,000 − 30,000 = $120,000
>
> If the bank is charging 2 points on the mortgage, the closing costs will include a fee in the amount of 2% of $120,000, or $2400.

4. The Internal Revenue Service allows mortgage points to be deducted. This includes not only the points a homebuyer pays to obtain a mortgage (or to buy down the interest rate of a mortgage), but even to points paid on behalf of the buyer by the seller. The rule is retroactive to the year 1991. Regulations also stipulate that points paid on a loan to remodel a home are deductible in the year they are paid. However, points paid on a refinanced mortgage must be deducted gradually over the life of the loan.

5. Most financial advisers recommend against paying points unless the seller is willing to pay them, since the money could be better invested in an IRA or mutual fund. The principal advantage of buying down the mortgage rate is that it allows the buyer to qualify for a larger loan. At the same time, if qualifying is not an issue, and there is a choice between a loan with no points and a lower rate loan *with* points, the tax deduction might be a consideration. The lower rate will entail lower interest payments, and consequently a smaller annual tax deduction. But the points will be deductible in the year they are paid.

6. While there is no set formula, it is generally the case that one point of prepaid interest will buy down the mortgage rate by ¼%.

> ◆ A $100,000 mortgage at 8¼% per year carries a monthly payment of $751.27. The same loan at 8% costs $733.76 per month. The monthly difference is $17.51, which amounts to $210.12 over a year. This should be weighed against the single point of interest, amounting to $1000, which would be charged up front in order to get the lower rate.

7. The *contract interest rate* of a mortgage is the rate stated on the commitment letter and mortgage contract. This figure can be misleading.

More useful is the annual percentage rate *(APR)*, the effective rate of interest the borrower will pay each year. This rate is always higher than the contract interest rate because it not only incorporates the effect of compounding (which is usually semiannual) but also includes costs in the form of points, private mortgage insurance, or other fees. For fixed-rate mortgages, the APR is calculated over the life of the loan. For adjustable-rate mortgages, the APR is calculated over the period leading up to the first adjustment of the rate.

The formula for determining APR is complicated and beyond the resources of the typical consumer. However, all lenders are required to state the APR of every loan they offer (not just mortgages). All a buyer has to do is compare the APR of the loans that are available, rather than the advertised contract rates.

♦ A bank offers two fixed-rate loans, one at 8¾% and the other at 8¼%. The APR for the first loan is 9.125% and for the second is 9.3%. Which one is better will depend upon the situation. The buyer might need a lower contract rate in order to qualify for the loan. The seller might be willing to pay the points at the lower rate, which would reduce the APR.

QUALIFYING FOR A MORTGAGE

In a buyer's market such as the real estate market of the 1990s, banks, credit unions, and mortgage companies actively pursue customers by matching their competitors' best rates and sometimes offering to pay some of the closing costs. They will often engage in creative accounting so that their customers can qualify. In the past, a mortgage applicant had to meet strict cost-to-income ratios. Some banks have relaxed these standards, but the old ratios still provide a good test of what a prospective buyer can afford.

Many banks, especially large ones, sell their mortgages on the secondary market, which consists of investors who buy mortgages in bundles while retaining the issuing bank as the servicer of the loan. This type of transaction is regulated by the Federal National Mortgage Association (Fannie Mae), which dictates what types of loans may be resold. Their long-standing rule requires that a buyer should have a monthly-cost-to-monthly-income ratio of at most 28%. That is, the monthly cost of the house—which is the sum of the mortgage payment, property tax, and insurance premium—should not exceed 28% of the buyer's gross monthly income. This is called the *housing ratio*.

housing ratio

$$= \frac{\text{mortgage payment} + \text{property tax} + \text{insurance premium}}{\text{gross monthly income}}$$

The costs listed in the numerator are monthly costs. In order to determine how big a mortgage he or she can afford, a buyer can use this formula:

$$\text{lowest qualifying monthly income} = \frac{\text{total monthly housing cost}}{0.28}$$

Lenders who do not sell their loans on the secondary market may allow up to a 33% ratio, which considerably reduces the minimum qualifying income.

Table 6.7 shows the annual income level required to qualify for a 30-year mortgage in the amount of $100,000. It compares a 28% qualifying ratio with the 33% ratio some banks allow. The house is assumed to cost $125,000, and the buyer is putting 20% down. Property tax is estimated at 1.5% of the selling price, and insurance at 0.25% of the selling price. These are converted to monthly figures (by dividing by 12) and added to the monthly mortgage payment, which is taken from a mortgage table. If the monthly costs are divided by 0.28 (28%) or 0.33 (33%), the result is the minimum required monthly income at each qualifying ratio. This figure is multiplied by 12 to give the required annual income in the last two columns of the table.

Table 6.7. Income Required to Qualify for a Mortgage on a $125,000 Home

Rate	Cost per month	Lowest qualifying income (28% ratio)	Lowest qualifying income (33% ratio)
6%	$782	$33,510	$28,436
6.5%	$815	$34,900	$29,636
7%	$848	$36,325	$30,836
7.5%	$882	$37,780	$32,073
8%	$916	$39,260	$33,310
8.5%	$951	$40,765	$34,580
9%	$987	$42,300	$35,890
9.5%	$1023	$43,840	$37,200
10%	$1060	$45,420	$38,545

Another qualifying ratio is the *total obligation ratio*. This is the ratio of *all* monthly obligations—mortgage plus car loan plus personal loans plus alimony, for example—to gross monthly income. The Fannie Mae limit is 36%, although some lenders will allow up to 38%.

total obligation ratio

$$= \frac{\text{total monthly housing costs} + \text{other monthly debt}}{\text{gross monthly income}}$$

$$\text{lowest qualifying monthly income} = \frac{\text{total monthly obligation}}{0.36}$$

• A mortgage applicant wishes to buy the $125,000 house referred to in Table 6.7. At a fixed rate of 8½%, the total monthly cost would be $951. If the applicant also has $300 per month in other payments (car loan, credit cards, etc.), her total monthly obligation is $1251. To find the minimum qualifying gross monthly income, divide the total monthly obligation by 0.36.

$$\$1251 \div 0.36 = \$3475$$

To find the minimum qualifying annual gross income, multiply this figure by 12.

$$\$3475 \times 12 = \$41{,}700$$

Not only does a buyer have to qualify for the mortgage, but the property has to qualify as well. If the bank believes that the property is not worth the selling price, it may reduce the amount of the loan or simply refuse to offer one. This determination is made based on the *loan-to-value (LTV) ratio*.

$$\text{loan-to-value ratio} = \frac{\text{loan amount}}{\text{appraised value of property}}$$

The maximum LTV ratio for a mortgage on a one- to four-family owner-occupied residence is 90%. Thus if a buyer makes an offer on a house which is then appraised at a lower value, the bank may ask for additional collateral or require the buyer to purchase private mortgage insurance.

• A house sells for $125,000 but is appraised at $110,000. If the buyer wanted to put down 20%, the loan amount would be 80% of $125,000, which is $100,000. The loan-to-value ratio would be

$$\frac{\$100{,}000}{\$110{,}000} = 0.909 = 90.9\%$$

Because this exceeds the limit, the bank would ask for a larger down payment. If the buyer put up another $1000, the ratio would be satisfied.

$$\frac{\$99{,}000}{\$110{,}000} = 0.90 = 90\%$$

PROPERTY TAXES

Most cities and towns generate revenue for public services by levying a *property tax*. The tax is stated as a dollar amount per $1000 of assessed value. The *assessed value* is based on such factors as a house's square footage, its condition, the lot size, and the value of surrounding properties. The assessed value is usually not the market value. In some localities, the assessed value is calculated directly as a percent of the market value—perhaps 80%.

- If a house sold for $140,000 and is assessed at 80% of market value, its assessed value would be ($140,000) × (0.8) = $112,000. If the tax rate is $15 per $1000, the annual property tax would be

$$(\$15) \times (112) = \$1680 \text{ per year}$$

Property-tax bills are generally collected quarterly or semiannually rather than monthly. Often a bank holding a mortgage establishes an escrow account for the purpose of paying property taxes; the accumulated three or six months of payments are paid to the city or town by the bank at the end of each billing period, which is based on the fiscal year (typically ending September 30).

FIXED-RATE MORTGAGES

Until the 1980s the standard mortgage was a 30-year fixed rate mortgage. But as interest rates rose precipitously in the early 1980s, fixed rates became unattractive and unaffordable. In response, banks came up with adjustable-rate mortgages, which allowed buyers to meet qualification ratios by establishing lower initial rates of interest.

In a *fixed-rate mortgage,* the annual rate remains the same for the life of the loan. Consequently, the monthly payment also remains the same. At the outset of the loan, these payments consist mostly of interest, with relatively little of the principal being paid off. But with each payment this balance changes; the portion constituting interest diminishes while the retirement of principle increases. This is described in detail in the section on amortization schedules (see page 276).

ADJUSTABLE-RATE (VARIABLE-RATE) MORTGAGES

The advantage of an *adjustable-rate mortgage (ARM)* or *variable-rate mortgage (VRM)* is that it offers an initial rate that is well below the market rate for fixed-rate loans. Therefore buyers can qualify more easily and enjoy lower payments in the initial period of the loan. The disadvantage of ARMs is their uncertainty. At regular intervals, the rate is adjusted to reflect the prevailing interest rates in the securities markets. An adjustable rate can jump as much as 2 to 3 percentage points at once, and the monthly payment can therefore increase considerably. Many buyers of ARMs are surprised to find that their rate has risen over an adjustment period in which national rates have fallen. This results from a misunderstanding of the terms of ARMs.

1. The Initial Interest Rate ARMs are offered at tempting rates that entice many buyers who would not qualify for a fixed-rate loan. This *initial rate*, or *teaser rate*, can be misleading. When the adjustment period comes up, the discounted teaser rate is readjusted according to an index.

2. The Index Banks set interest rates on ARMs according to a formula that is based on a key economic *index*, one of the market indexes that can be found each week in business newspapers. The most commonly used indexes are:

> the *contract rate:* a nationwide average of fixed-rate mortgage rates.
> the *cost-of-funds index:* a national average of the interest rates banks and S&Ls pay to borrow money.
> *treasury yields:* the average rates of return on U.S. government securities. (A bank may use a specific rate such as the six-month Treasury-bill rate.)
> the *prime rate:* the rate of interest charged by the biggest banks for their best customers.
> the *LIBOR* (London Interbank Offered Rate): a kind of international prime rate.

3. The Margin A *margin* is a number that is added to the index in order to arrive at the *fully indexed rate*. It is important to ask what each lender's margin is, because this number determines how the rate will be set at each adjustment.

> ♦ A one-year ARM is offered at 6½% per year. The margin is 2½% and the index is the six-month Treasury-bill rate. If in the first year the T-bill rate falls from 5¼ to 4¼%, the ARM rate would still rise to 4¼ + 2½ = 6¾%.

4. Caps An *interest rate cap* determines how much the interest rate on an ARM can change at each adjustment period. Most ARMs carry an adjustment cap of around 2%. The mortgage should also have a *life-of-loan cap*, which defines a maximum interest rate that cannot be exceeded. A typical life-of-loan cap is 5–6% over the initial rate.

- A one-year ARM with an initial rate of 6½% per year and a "2 and 5" cap can rise 2 percentage points each year, up to a maximum increase of 5%. The worst-case scenario would be a steady rise in interest rates that results in rates of 6½%, 8½%, 10½%, and finally 11½% through the first four years of the loan.

ADJUSTABLE-RATE, FIXED-PAYMENT MORTGAGES

One type of adjustable-rate mortgage is a loan with a fixed monthly payment and a variable amortization. If the interest rate changes at the end of an adjustment period, the amortization schedule changes—that is, the part of the monthly payment that constitutes interest will change. Thus the rate at which the unpaid balance is retired can increase or decrease from adjustment period to adjustment period.

The danger of a fixed-payment loan is that, if the interest rate goes too high, the monthly payment will not be enough to cover the monthly interest charge. When this happens, the unpaid interest is charged to the principal, so the amount owed (the unpaid balance) actually increases with each monthly payment. This is referred to as *negative amortization*.

BALLOON MORTGAGES

A *balloon mortgage* is a loan that is broken into a series of installment payments followed by a single large payment. (A *balloon payment* is a payment—often the entire original loan amount—that comes due after a period of relatively modest monthly payments.) The appeal of a balloon mortgage is the monthly installment, which is smaller than the monthly payments for fixed and adjustable rate mortgages—for a while.

Balloon mortgages work by allowing the borrower to pay only interest (or mostly interest) on the principal for an initial period. The disadvantage is that the borrower builds little or no equity. In one type of balloon mortgage, none of the principal is retired by the installment payments. This type of loan is typically contracted over a three-to-five-year period. Most borrowers in this situation end up refinancing the mortgage when the balloon payment comes due. In another type of balloon mortgage, part of the loan has been repaid after the initial period, but a significant sum is due at maturity, which might be 15 years.

REFINANCING

When interest rates fall to record low levels, holders of fixed-rate mortgages and ARMs are tempted to refinance their loans at lower rates. But because refinancing incurs most of the closing costs paid at the original settlement, the common wisdom has held that refinancing is only worthwhile when the interest-rate differential is at least 2%—that is, when it will result in a rate reduction of 2% or more.

Some analysts now say that even a 1% reduction justifies the switch, but the decision rests on more factors than a simple formula can express. The decision to refinance should take into account the number of years the owner will keep the house, the amount of the tax deduction, and the yield on the owner's other investments. After all, buying a house should be thought of as an investment, one that ties up money that could otherwise earn interest elsewhere. The value of this equity, the tax advantages of a mortgage, and the cost of financing the property should be assessed in the context of the homeowner's entire investment picture.

Investing Money

The goal of investing is to increase the future spending power of money. People invest in order to earn money that will offset the effects of inflation. Very successful investors earn a living on investments alone.

All investment decisions pit two competing factors against each other: the rate of return and the amount of risk. In the world of investments, as in many areas of life, the greater the risk, the higher the potential return. The most secure monetary investments, considered risk-free, are U.S. government securities such as Treasury bills, notes, and savings bonds. Historically, these have paid rates of interest that keep pace with inflation. But when the tax on interest earnings is taken into account, these investments have historically fallen behind the cost of living. The same can be said for many other guaranteed investments such as passbook savings accounts, money market accounts, and municipal bonds. Consequently, an investor who is looking to beat the inflation rate must take some risks. The risks can be balanced, however, if investments are diversified. To do this, the investor designs a portfolio, which is a plan reflecting an investment philosophy that pits possible gains against acceptable risks. The portfolio might mix stocks, bonds, government securities, mutual funds, and other investments in a way that takes advantage of tax deferments, tax deductions, the prospect of steady income, or the possibility of steady growth.

The purpose of this and the following sections is to sketch the mathematics of investments and thereby provide a foothold on an inexact science, but one that clearly favors the prepared mind.

SECURITIES

When a privately owned company reaches a stage of its growth where it wants to raise money (referred to as *capital*), it may go public by offering to sell shares of ownership. These shares take the form of *stock,* which is then sold (or traded) on the floor of a stock exchange. A broker acts as a buyer for investors, and facilitates their transactions.

Another way for a corporation or government to raise money is to issue *bonds,* which are promissory notes that pay interest in the form of dividends. With stock, ownership of the corporation is actually conferred to the stockholders as a group, who then share in the profits and losses. With bonds there is no conferral of ownership, and the risk is reduced, although not eliminated. When a company goes bankrupt, any money that remains is distributed first among the bondholders and then among the stockholders.

For those who prefer to leave investment decisions to a professional, *mutual funds* offer a way to invest in a wide selection of bonds or stocks. Investment companies make buying shares in a mutual fund as simple as opening a bank account.

Stocks, bonds, mutual funds, legal notes, certificates of deposit (CD's), and debentures are all forms of *securities,* written documents of ownership or creditorship used to establish an investment in some business enterprise. All securities are ultimately judged by the amount of money they earn for the investor calculated as a percent of the amount invested. This is expressed as a *yield,* which relates earnings to price. Yields, earnings, dividends, and other investment terms are defined below.

KEY ECONOMIC INDICATORS

An investor's rate of return is always measured against indexes that show how the securities market has performed as a whole. These indexes are constructed by averaging the performance of a selected group of securities. The actual index number has meaning only in reference to index numbers from other periods. That is, index numbers from year to year form ratios that represent the percent growth of the market.

$$\text{percent change of an index} = \frac{\text{last value} - \text{initial value}}{\text{initial value}} = \frac{\text{change}}{\text{initial value}}$$

1. The *Dow Jones Industrial Average* measures the performance of 30 *blue-chip stocks* that are traded on the New York Stock Exchange. These are stocks issued by companies whose economic stability is thought to reflect the state of the national economy. Thus "the Dow" is a gauge of the stock market as a whole, and to "beat the Dow" is a standard measure of a portfolio's annual performance.

Until recently, any change over 100 points in the Dow would have indicated that some serious news had shaken the market. This is no longer the case. The Dow first broke the 3000 mark in 1991, the 5000 level in 1995, and the 8000 level in 1997. Thus a 100-point gain or loss has diminished from a 3.3% to a 2% to a 1.25% change.

Table 6.8 gives the Dow Jones average for each quarter of 1991 in order to show how rates of growth are calculated. A comparison of the September 30 average of 3016.77 and the December 31 average of 3168.83 shows an increase of 152.06 points. The ratio of 152.06 to the initial average of 3016.77 gives the percent change of 5.04%. If this three-month figure were used to calculate the annual growth, it would be multiplied by 4, which results in 20.16%. The actual annual rate of growth is found by calculating the change over the entire year (535.17 points), and then dividing by the initial average (2633.66 points), which gives an overall percent increase for 1991 of 20.32%.

Table 6.8. Quarterly Dow Jones Industrial Averages for 1991

Date	Closing average	Quarterly change	Percent change
Jan. 1	2633.66		
March 28	2913.86	+280.20	+10.64
June 28	2906.75	−7.11	−0.24
Sept. 30	3016.77	+110.02	+3.78
Dec. 31	3168.83	+152.06	+5.04

2. The *Standard & Poor 500 Index* is similar to the Dow Jones average but is based on 500 stocks that are spread across a wide range. Like the Dow, it includes industrials, in addition to financials, transportation stocks, and utility companies.

There are many other indexes that go beyond the Dow Jones or Standard & Poor 500 by focusing on narrowly defined categories or on specific commodities exchanges such as NASDAQ (see below). These work like all indexes: they can be used to chart the percent change of a class of investments from day to day, month to month, or year to year.

3. The *Gross National Product (GNP)* and the *Gross Domestic Product (GDP)* are two measures of economic activity in the United States. As such they are seen as indicators of economic growth.

The Gross National Product is the total cost of all goods produced by labor or property that is supplied or owned by residents of the United States. This includes goods produced in foreign countries by or for American business interests.

The Gross Domestic Product is a similar measure of the cost of goods produced, but it includes only those goods produced *within* the United States proper by any U.S. residents. Thus the GNP is a measure of the level of total U.S. production, while the GDP focuses only on domestic production. The GDP is currently viewed as a better indicator of the national economy than the GNP, since it more accurately reflects the levels of unemployment, productivity, and industrial activity. The GNP, by contrast, is seen as a better measure of purely economic activity of American businesses, many of which may be producing goods abroad. At this time, the numerical difference in GNP and GDP is slight.

Stocks

The price of a share of stock is determined by market mechanisms at work on the trading floors of stock exchanges. There are two major stock exchanges in the United States, the New York Stock Exchange (NYSE) and the American Stock Exchange (AMEX). There is also a national market for *over-the-counter stocks*. This is not an actual trading floor as much as a virtual marketplace conducted over telephones, wire services, and computer networks. This trading information is tabulated by the National Association of Securities Dealers, whose computerized quote system is called NASDAQ (National Association of Securities Dealers Automated Quotations). In addition to these, there are regional stock and commodities markets in major cities such as Boston, Chicago, Philadelphia, and Los Angeles, as well as many overseas stock exchanges.

Anyone may purchase stock by getting in touch with a stockbroker, who will arrange to buy or sell stocks in return for a commission. This charge will be either a flat fee, a sliding-scale fee, or a percentage of the transaction. Almost all the information relevant to the performance of a stock can be found in major daily newspapers.

READING STOCK- EXCHANGE LISTINGS

Stock tables consist of columns that identify a stock and give the most relevant information about its performance. A typical listing for two stocks might look like this:

| 52-week | | | | Yld | | Sales | | | | |
High	Low	Stock	Div	%	P/E	100s	High	Low	Close	Change
32¼	20⅞	AbCo	.56	2.3	10	237	25⅜	23	24⅛	+⅛
30⅜	26½	AbCo pf	3.16	10.7	. . .	168	30⅛	29½	29⅝	+⅛

Key:

High/Low: 52-week high and low. Stock prices are given as *points*, which are dollar amounts per share. The high and the low prices listed in the first two columns give the highest and lowest prices the stock has been traded for within the last 52-week period. Dollar amounts are usually broken down no further than eighths, although increments for low-priced stocks may be as small as thirty-seconds. One-eighth of a point is equivalent to 12.5 cents; one thirty-second of a point is 3.125 cents.

Stock: This column identifies an issue of stock by the company name and the type of stock. It may also give some information about the stock's history or status. The abbreviations of stock names come from the Associated Press market quotations supplied daily to newspapers.

The two types of stock are *common stock* and *preferred stock.* Preferred stock, identified by the letters *pf* next to the company name, pays a dividend each year (usually in four installments) at a set percentage rate.

Div: This column indicates the estimated yearly dividend paid based on the dividends that have been paid thus far. For AbCo's common stock, the dividend is 56 cents per share. AbCo preferred stock is paying $3.16 per share annually.

Yld %: The *yield percentage* is the percent return on investment as calculated from the day's closing price. Specifically, it is the ratio (expressed as a percent) of annual earnings (dividend per share) to closing price. For AbCo preferred,

$$\text{Yld\%} = \frac{\text{dividend}}{\text{closing price}} \times 100\% = \frac{3.16}{29.625} \times 100\% = 10.667\% \approx 10.7\%$$

P/E: The *price/earnings (P/E) ratio* expresses the ratio of the day's closing price divided by per-share earnings. The earnings per share are found in the company's quarterly reports and are calculated as the earnings over the last four quarters. P/E ratios are not given for preferred stock.

Sales 100s: This is the volume of shares traded in that day, given in units of hundreds. On this day, 23,700 shares of AbCo common stock changed hands.

High: The highest price at which the stock was traded that day.

Low: The lowest price at which the stock was traded that day.

Close: The price of the stock at the last sale of the day.

Change: The net change in selling price, calculated as the difference in closing prices between the two previous days.

PRICE/EARNINGS RATIO

The *price/earnings ratio,* also known as the *multiple,* is used by investors to judge the earning potential of a stock. It can be calculated in two ways.

The *trailing P/E* is the ratio of current price per share to earnings, where the earnings are taken from the previous year's financial report.

The *forward P/E* is the same ratio calculated from the projected earnings over the coming year.

Earnings per share are calculated as follows:

$$\text{earnings per share} = \frac{\text{profit after taxes on most recent financial report}}{\text{total no. of shares outstanding}}$$

◆ A stock selling for $40 with a previous year's earnings of $2 per share has a P/E ratio of 20. If projected earnings for the next year are $4 per share, the forward P/E would be 10.

There is no recommended range of P/E values; each company's P/E ratio must be judged in comparison to those of companies in the same industry. Table 6.9 shows P/E averages for various industries, listed from high to low. Stocks within any of these categories can be judged against the average in the following way: A low P/E indicates a mature, steady-growth company. Blue-chip stocks in particular are noted for paying regular dividends and maintaining good earnings per share. High P/E stocks have lower annual yields and are often fast-growing companies that reinvest most of their profits back into the company. These tend to be riskier, aggressive growth stocks.

P/E ratios are inversely related to interest rates. This means that if interest rates rise, P/E numbers tend to fall, and vice versa. When a stock is described as being *undervalued*, it means that its P/E seems low compared to the industry average.

STOCK SPLITS

As the value of a company's stock grows, the board of directors may decide to *split* the stock. When this occurs, each share converts to some multiple of shares and the price is adjusted accordingly. A typical split is a *two-for-one*, in which the number of shares doubles and the price per share is halved. Another split is a *three-for-two*, in which the price decreases by one-third.

The reason for a split is largely psychological. It is based on the idea that a price of $30 or $40 a share is more attractive and accessible than $60 or $80 a share. Stocks that have climbed rapidly to the $70–90 range are considered ripe for a split.

What experienced investors look for is not the fact of a split but the effect on the price/earnings ratio. In general, the lower the P/E the better.

MARGIN ACCOUNTS

A *margin account* offers a way to purchase stock without paying for it in full up front. The purchaser who buys "on margin" puts up only a frac-

Table 6.9. Price/Earnings Ratios by Industry in a Recent Year

Industry	P/E ratio	Yield
Publishing/ Broadcasting	64	1.67
Beverages	39	1.80
Fuel	35	3.03
Software	29	0.37
Manufacturing	29	1.84
Paper/Lumber	27	2.70
Transportation	27	1.43
Leisure industries	26	2.75
Chemicals	24	2.75
Service industries	23	1.38
Automotive	22	1.84
Retail	22	1.10
Medical products	22	1.39
Drug retail/distribution	21	1.44
Housing/Real estate	21	1.54
Home furnishings	20	1.80
Telecommunications	20	3.90
Aerospace	19	3.06
Drugs and research	19	2.60
Electrical/Electronics	19	1.70
Food	19	1.85
Health care services	18	0.67
Containers/Packaging	17	1.60
Conglomerates	16	2.78
Non-bank financial	16	2.46
Computers/Peripherals	15	0.65
Tobacco	15	2.33
Banks	13	2.82
Utilities and power	5	5.30

[Source: *Business Week*]

tion of the purchase price, and a broker lends the remaining amount. The broker then holds the stock as collateral for the loan.

The *margin* is the percentage of the purchase price that has been paid by the purchaser.

$$\text{margin} = \frac{\text{market value of stock} - \text{amount of loan}}{\text{market value of stock}}$$

This should be expressed as a percent.

• A buyer pays $2500 for 100 shares of stock at $40 per share. The market value of the stock is $4000. To make up the difference, the buyer takes a $1500 loan from the broker.

$$\text{margin} = \frac{4000 - 1500}{4000} = 0.625 = 62.5\%$$

This percentage is called the *initial margin.* The smallest allowable initial margin is 50%. Once established, the margin will change as the value of the stock changes. Any revised value of the margin is called a *maintenance margin.* On the New York Stock Exchange, a maintenance margin is not allowed to fall below 25% (some brokers set an even higher limit). When the margin does threaten to fall below the minimum, the stock is referred to as *under-margined,* and the buyer will be asked to come up with more cash in order to reduce the size of the loan, and thereby increase the maintenance margin. This is a *margin call.* If it is not met, the broker will immediately sell the stock to cover his loan.

• If the stock in the example above fell to $20 per share, the market value of 100 shares would fall to $2000, and the margin would fall to

$$\frac{2000 - 1500}{2000} = 0.25 = 25\%$$

The broker would issue a margin call at this point. If forced to sell, the broker would recoup his $1500 loan plus any commissions or fees that are owed. The remainder would go to the buyer.

SHORT SELLING

Short selling is a way of profiting from the decline in value of a stock. The buyer, anticipating a drop in a stock's price, uses a broker to borrow stock belonging to someone else. The stock is then sold, and the proceeds are held by the broker. At a later time, the borrowed shares must be repurchased and returned. Any dividends the stock earned during the intervening time must be paid to the owner of the stock. If the price falls during this time, the short seller can keep the difference between the original selling and final repurchase price, minus any dividends. He must still pay a commission, both when the stock is sold and again when it is repurchased.

The short seller is required to put up a percentage of the value of the borrowed stock, usually 50% of its value. This is called a margin.

Short selling carries a high level of risk and can result in a substantial loss if the price of the stock goes up. Securities regulations also stipulate that a stock cannot be sold short while its price is falling; the price must first rise before the stock can be sold.

Bonds

A *bond* is a debt security issued by a company, a municipal government, a federal agency, or the federal government itself. As a means of raising money, it is essentially a loan that pays interest in a variety of forms.

A bond has a *face value* (or *par value*) and a maturity date at which it can be cashed in. The face value for most bonds is $1000, although there exist so-called *baby bonds* that have denominations of $500 or less. The maturity period is usually 15 years, 20 years, or a higher multiple of 10 (up to 100 years, for what is known as a *century bond*). Most bonds also have stated rates of interest, called *coupon rates* because at one time many corporate bonds had tear-off coupons that could be redeemed periodically. The coupon rate does not change during the life of the bond

Bonds are rarely sold for par value. Instead they are sold at either a discount or a premium. A *premium* is the amount above par value at which a bond sells; a *discount* is the amount below par.

Because their dividend rates are fixed, the value of bonds fluctuates with market interest rates. When interest rates go down, bonds increase in value. When interest rates go up, bonds decrease in value. A bond's market value will also change (decrease, usually) as it gets closer to maturity.

READING BOND LISTINGS

Daily newspapers carry listings for the stock and bond markets in the format shown below. These statistics summarize the previous day's trading activity. A fictitious company's bond listing might look like this:

Bonds	Cur. Yld.	Vol.	High	Low	Close	Net Chg.
AbCo 12½s99	12	46	101¼	99¾	100⅞	+¼

Key:

Bonds: The first column lists the name of the bond (using Associated Press abbreviations), its *coupon interest rate* (12½% per year, usually paid semiannually), and its *maturity date* (in this example, September 1999). The interest rates are listed in increments of ⅛%.

Cur. Yld.: The *current yield* is the rate of return that the buyer would earn by buying at the closing price.

$$\text{current yield} = \frac{\text{annual interest earned}}{\text{cost per bond} + \text{commission}}$$

The annual interest payment is the stated rate (12½%) times $1000.

If this column contains the letters *CV*, then the bond is *convertible* into stock.

Vol: The volume is the total number of bonds traded on the previous day.

High, Low, Close: These three columns give the highest, lowest, and last price paid in the previous day's trading. These figures must be multiplied by 10 to find the actual dollar value. Here, the high price was 101¼ × 10, or $1012.50.

Net Chg: This is the *net change* in price from the previous day's closing price. To convert to dollars, convert the figure to a decimal and multiply by $10. A change of +¼ means an increase of 0.25 × $10, or $2.50 per bond from the previous day.

YIELD-TO-MATURITY

The interest paid on a bond is based on its par value, not its selling price. But its listed interest rate is not the best indicator of the bond's performance. More revealing is the *yield-to-maturity,* which can be compared to prevailing market interest rates to determine the true market value of a bond.

Yield-to-maturity can be calculated for a bond by following these steps:

i. Find the amount of the discount (D) or premium (P).

ii. Divide D or P by the number of years to maturity. This is called the *annual discount* or *annual premium.*

iii. If working with a premium, deduct the annual premium from the annual interest earned. If working with a discount, add the annual discount to the annual interest earned. This new figure is the *adjusted annual interest.*

iv. Find the average of the original cost of the bond and the anticipated proceeds from its sale using this formula:

$$\frac{(\text{price} + \text{sales commission}) + (\text{face value} - \text{sales commission})}{2}$$

v. The yield-to-maturity is found by dividing the result of step 3 by the result of step 4.

♦ A 9% bond maturing in five years is purchased at 108 (which means $1080—the premium is $80). Assuming a $5 commission, the total cost is $1080 + 5 = $1085. The annual interest is 9% of $1000, which is $90 per year. Assume also that the commission will be $6 when the bond is sold. The yield-to-maturity is found by dividing the premium of $80 by 5 (years) and deducting the result from the annual interest of $90 (steps 2 and 3).

$$90 - \frac{80}{5} = 90 - 16 = \$74$$

The cost of purchase was $1085. The selling price will be $1000 minus the $6 commission, or $994. The average of $1085 and $994 is $1039.50. Therefore the yield-to-maturity is

$$\frac{74}{1039.50} \approx .0712, \text{ or } 7.12\%$$

U.S. Government Securities

The federal government raises money by issuing securities in the form of *Treasury bonds, Treasury bills,* and *Treasury notes.* The distinction is a matter of years to maturity. Treasury bonds are issued for terms from five to 30 years; the most popular is the 30-year bond. Notes are issued in terms from five to seven years. Treasury bills mature in a year or less; the most common T-bill maturities are three months, six months, nine months, and one year.

TREASURY BONDS AND NOTES

1. Treasury bonds and notes are issued in denominations of $1000. They pay interest twice a year. The *face value* of the bond is its value at maturity, and its *coupon rate* is the annual rate of interest that it pays.

2. A $1000 bond has a face value of $1000, but this may not be its sale price. Treasury bonds and notes are typically bought at either a discount (a price below face value) or at a premium (a price above face value). The difference between the purchase price and the face value is referred to either as the discount or the premium.

◆ A $1000 bond purchased at $850 has a discount of $150. It can also be described as "discounted $150."

The market prices of bonds and notes change week by week in response to changes in interest rates and other economic indicators. Thus a bond or note has three prices: a face value, a purchase price, and a current market value. The market value depends on the coupon rate and the years to maturity.

3. The semiannual interest payment for a bond or note is calculated using the face value, not the purchase price. It is based on the coupon rate, which is specified on the bond. It is calculated as simple interest using a 365-day year.

* A $1000 bond with a coupon rate of 8% earns a semiannual interest payment of

$$\frac{1}{2} \times (\$1000 \times .08) = \$40$$

4. Bond and note purchase prices fluctuate from week to week, but the interest they pay is always calculated on their fixed face value. Consequently it is not immediately apparent what rate of return is being earned on the money invested (which usually does not coincide with the face value). The measure of return on investment is called the current yield (see page 296). It is calculated as follows:

$$\text{current yield} = \frac{\$1000 \times \text{coupon rate}}{\text{purchase price}}$$

* A $1000 bond with a 9% coupon rate purchased for $875 has a current yield of

$$\frac{\$1000 \times .09}{\$875} \approx 0.1029 = 10.29\%$$

This percent represents the interest rate of return on the $875 investment.

5. Another measure of the value of a government bond is its yield-to-maturity (YTM), discussed above with respect to corporate bonds. This is a rate of return that takes into account the number of years until maturity, the interest earned over those years, and the gain or loss when the bond is cashed in at maturity. It is a measurement of how the bond "performs" over the years it takes to mature.

YTM figures for government bonds and notes are given in published tables. The calculation is complex, but the approximate figures can be found with a simple formula, as follows:
Let

$$i = \text{interest earned per year} = \$1000 \times \text{coupon rate}$$

$$a = \frac{\$1000 - \text{current market price}}{\text{years to maturity}}$$

$$b = \frac{\$1000 + \text{current market price}}{2}$$

Then

$$\text{yield-to-maturity} = \frac{i + a}{b}$$

• Assume a $1000 bond with a 9.5% coupon rate and 20 years remaining to maturity has a market value of $1150. Its approximate yield to maturity is calculated as follows:

$$i = \$1000 \times .095 = \$95$$

$$a = \frac{\$1000 - 1150}{20} = -7.5$$

$$b = \frac{\$1000 + 1150}{2} = \$1075$$

$$\text{YTM} = \frac{\$95 - \$7.50}{\$1075} = 0.0814 = 8.14\%$$

READING TREASURY BOND AND NOTE QUOTATIONS

The business pages of most large daily newspapers list price and yield quotations for U.S. government securities. The listing reprinted from the *New York Times* below is typical.

Month	Rate	Bid	Ask	Chg	Yld
Apr98 n	7.875	105-10	105-14	+01	5.42
Apr98 n	5.125	99-11	99-15	. . .	5.42
May98 n	6.125	101-18	101-22	−01	5.43

Key:

Month: The date gives the month and year of maturity. The letter *n* indicates that it is a note; the absence of a letter indicates a bond.

Rate: The rate is the coupon rate of interest. Rates are listed in increments of one-eighth of a percent. This is the percent used to calculate the semiannual interest payment.

Bid: The *bid price* is the price at which the note or bond is offered to dealers. All prices are given as a percent of the face value, which is $1000. However, fractions of a percent are given as multiples of $\frac{1}{32}$. Thus 105–10 means $105\frac{10}{32}\%$. The decimal form of $\frac{1}{32}$ is 0.03125, so $\frac{10}{32}$ equals 0.3125, and $105\frac{10}{32}\%$ converts to 105.3125%. Move the decimal point two places to the left and the percent becomes a decimal multiplier. The note sells for $1.053125 \times \$1000 = \$1053.125$.

Ask: The *ask price* is the price at which a dealer will sell to an investor. The price is given as a percentage, which can be converted in the same way as the bid price. 105–14 means $105\frac{14}{32}\%$ of $1000. This converts to 105.4375% of $1000, or $1.054375 \times \$1000 = \$1054.375$.

Chg: The *change* in price from the previous day's closing price is also in units of $\frac{1}{32}\%$.

Yld: This is the *yield to maturity* as calculated from the ask price.

U.S. TREASURY BILLS (T-BILLS)

1. *Treasury bills* are debt obligations that mature in a year or less. The most common T-bills mature in three months, six months, or one year. They are sold in denominations starting at $10,000.

2. Unlike treasury bonds or notes, T-bills pay *discount interest* (see page 262). That is, they are sold at a discounted price and can be redeemed for their face value at maturity. The discount is calculated as a straight percentage using a 360-day year and a stated rate called the *discount rate*.

* A 91-day (13-week) T-bill is discounted at 9%. The buyer would pay $10,000 minus the discount interest, where

$$\text{discount interest} = \left(\$10,000 \times .09 \times \frac{91}{360}\right) = \$227.50$$

Thus the buyer pays $10,000 − 227.50 = $9772.50. The discount is not compounded. It is simple interest which the buyer gets up front through the discounted price, but may not realize until the bill is redeemed for its full value.

3. Because interest on T-bills is paid in a fundamentally different way than interest on treasury bonds or notes, some method of comparing their rates of return is needed. This is called the *bond equivalent yield*. It reflects the fact that T-bill interest is calculated over a 360-day year, while notes and bonds use a 365-day year.

$$\text{bond equivalent yield} = \frac{365 \times \text{discount rate}}{360 - (\text{discount rate} \times \text{days to maturity})}$$

The T-bill in the previous example has a discount rate of 9% and 91 days to maturity. Its bond equivalent yield is

$$\frac{365 \times .09}{360 - (.09 \times 91)} = 0.0934 = 9.34\%$$

READING TREASURY BILL QUOTATIONS

The *Wall Street Journal* gives daily quotations for government securities. A typical listing for T-bills might look like this:

Date	Days to Mat.	Bid	Asked	Chg.	Ask Yld.
Dec 21	7	5.58	5.56	−0.09	5.68
Dec 28	14	5.46	5.44	. . .	5.56
Jan 4	21	5.17	5.15	+0.05	5.27

Key:

Date: The date is the date of maturity.

Days to Mat.: The *days to maturity* is self-explanatory.

Bid: The *bid price* is a percent discount rate at which dealers can buy the bill. The actual dollar price is equal to $10,000 minus the discount, where discount = discount rate × 100 × days to maturity ÷ 360. For example, a listed discount rate of 5.29 on a 91-day T-bill would carry a discount of 5.29 × 100 × 91 ÷ 360 = $133.72. The dollar bid price would be $10,000 − $133.72 = $9866.28.

Asked: The *asked price* is the price at which the dealer will sell to a buyer. The price is given as a percent in the same way as the bid price.

Chg: The change in price from the previous day's trading is also given in percent increments. These are easily read as *basis points.* A basis point is $^1/_{100}$ of a percent, or .01%. A change of −0.09 is a drop of 9 basis points.

Ask Yld: This column gives the *bond equivalent yield* in percent form.

Mutual Funds

A *mutual fund* is a diversified investment that pools investors' money into a stock and/or bond portfolio. An investor can buy shares in a fund, and the value of each share fluctuates with the performance of the portfolio. Every fund has a manager who follows an investment strategy that is laid out in general terms in a prospectus. Some funds spread the money over as few as 20 stocks, others invest in up to 200. A fund may promote itself as aggressive or conservative. The most conservative, low-risk (and steady-yield) funds tend to buy corporate bonds and preferred stock only. On the other end of the scale, there are speculative funds that invest in progressive start-up firms.

Mutual funds constitute a large share of the investment market, since there is a mutual fund for every taste. There is also a basic vocabulary that must be mastered by anyone seriously interested in such investments.

1. Issuers of mutual funds can be either *open-end investment companies* or *closed-end investment companies.* The term *open-end* refers to an investment company that issues an unlimited number of shares to anyone who wants to buy in. In contrast, a *closed-end company* issues a fixed number of shares, which are listed on a major stock exchange. Once a closed-end company's shares have been issued and the initial sale is complete, a buyer must buy shares from a shareholder through a broker. Unlike an open-end investment fund, which sells directly to the investor, a closed-end company does not issue or redeem shares, and the price per share is determined on the open marketplace.

2. Investment companies pool investors' money in funds that are spread over stock and/or bond portfolios in accordance with an investment philosophy. The company may offer several funds, each with its own philosophy and manager. These are described in a publication called a *prospectus*. The investment company usually retains from ½% to 1% of the total investment as its annual *management fee*.

3. There are too many types of portfolios to list here, but they fall into some general categories. One way to categorize them is by investment philosophy. Here are the four most basic approaches:

> An *income fund* invests in corporate bonds and dividend-paying stocks that generate immediate dividend income for the investor.
> A *growth fund* targets stocks that have long-term growth potential but may not pay dividends or generate immediate revenue.
> A *balanced fund* seeks to split funds between income (dividend-paying, slow-growth) and growth stocks and bonds.
> A *performance fund* targets high-risk stocks with the potential for high yields.

Funds may also be classified by the type of securities they target. Here are five basic types (out of many):

> *Stock funds* invest the pool solely in common and preferred stock.
> *Bond funds* target bonds only.
> *Stock-and-bond funds* split the pool between stocks and bonds.
> *Municipal-bond funds* concentrate on municipal bonds with tax-exempt interest earnings.
> *Money-market funds* invest in a variety of short-term debt instruments such as Treasury bills.

4. Some mutual funds are purchased through securities dealers, brokers, or banks. Such a fund may carry a sales commission, which is referred to as a *load*. The load can be deducted from the investment either up front or when the shares are redeemed. An up-front charge is called a *front-end load*. A *back-end load fund* charges a commission when it is sold. A *no-load fund* is a mutual fund with no sales commission.

A typical commission is between 4¾% and 5¾% of the total investment. Thus when an investor spends $10,000 to buy into a 5% front-end-loaded fund, $500 goes to the broker or dealer, leaving $9500 worth of shares. The investor has to earn back $500 just to break even. By contrast, a no-load fund, which is not bought through a middleman, retains its full principal value.

Back-end-loaded funds often use a six-year declining load, which charges 5% if the fund is sold in the first year, 4% for the second, 3% for the third, 2% for the fourth, 1% for the fifth, and no load after the sixth. There are also *hybrid funds*, known as *low-load funds*. These charge a commission of about 3%.

5. An investor may buy into a mutual fund by purchasing a certain number of shares (where *total investment = no. of shares × price per share*). An alternative is to buy into the mutual fund at a lump-sum value which may not convert to a whole number of shares. (See page 311.)

READING MUTUAL-FUND QUOTATIONS

Listings of mutual-fund performance in the business pages of newspapers follow a standard format. The price of a share is determined by the *net asset value* (or *NAV*), which is a ratio of the investment fund's assets to the number of shares outstanding. Shares are sold at the NAV, to which sales commissions may be added. They are also redeemed at the NAV, minus any redemption charges.

Fund Family Fund Name	Type	Rating	NAV	Wkly % Ret	YTD % Ret	1-Yr % Ret	3-Yr % Ret
Acme Capital							
AcmeGro	G	5/2	18.75	–0.2	+23.6	+23.0	+12.1

Key:

Fund Family: Names of investment companies are given in boldface, followed by the individual funds they maintain. (The name given here is fictitious.)

Type: This listing follows the style of the *New York Times*. The key to abbreviations of bond types is given in a table included with the price quotations. In this instance, the *G* indicates that this is a growth fund.

Rating: The rating is a measure of risk vs. return as judged by Morningstar, Inc., a mutual-fund rating firm. The first number rates the return on investment. As explained by the *Times*, "a return rating of 5 shows top performance among similar funds over the last 3 years, while a rating of 1 shows poor performance. The ratings are based on overall return after deducting sales charges."

The second number assesses risk. It is based on the fund's downward performance as compared to Treasury bills during market downswings. That is, when interest rates rise, bond prices tend to fall. The benchmark for such downswings is the price of T-bills. Bonds that devalue more rapidly than T-bills are considered higher risk. On the scale of 1 to 5, 5 is the lowest risk rating, and 1 is the highest. Acme's ⁵⁄₂ rating shows strong performance but with relatively high risk.

The distribution of the ratings is such that about 10% of all rated funds fall into the extreme categories of 1 or 5.

NAV: The *net asset value* is the current price per share, not counting sales commissions. It is calculated by the following formula:

$$NAV = \frac{\text{total value of fund investments} - \text{debts}}{\text{no. of shares outstanding}}$$

Closed-end mutual funds are often discounted below net asset value.

Wkly % Ret: The total return for one week is a percent change in the net asset value of a share (the price per share). Any dividends earned on the securities in the portfolio are reinvested; therefore dividends show up in the calculation of net asset value.

YTD % Ret: This is the *year-to-date* return on investment.

1-Yr % Ret: This is the total percent increase in net asset value over the past year.

3-Yr % Ret: This is the total percent increase in net asset value over the last three years.

Tax-Deferred Investment Plans

A *tax-deferred investment* is a type of tax shelter that at one time was available only to the very rich. Changes in tax laws in the 1970s and 1980s, however, created a wide variety of tax-deferment plans that were available to the average wage earner. Among these, the *IRA* (or *individual retirement account*) is the best known and most popular. It is offered by almost every kind of financial institution—banks, credit unions, S&Ls, and investment companies.

The basic idea is very simple. The tax laws allow anyone to deposit up to $2000 per year out of their taxable income in an IRA account. For those in the lowest tax brackets, the IRA contribution is not taxed in the year it is earned. Instead, this pretax income is allowed to collect interest in an investment fund, and becomes available to the investor when he or she turns 59½. Taxes are then paid on the money in the year it is withdrawn. The assumption is that most retired people will be in a lower tax bracket and their IRA contributions will be taxed at a lower rate than during the years when they were set aside. More important, the untaxed income earns interest, which is not taxed until it is withdrawn.

Other tax-deferred plans—*SEP-IRA, Keogh,* and *401(k)*—operate on the same premise. Keoghs and SEPs, or *simplified employee pensions,* are intended for the self-employed or for employees of small firms, and 401(k)'s and 403(b)'s are available to employees of most public corporations, private companies, and nonprofit organizations such as universities and public schools. In each type of plan, the employee is allowed to set aside a percent of taxable income (which may be partially matched by the employer) before taxes are taken out. These amounts may exceed the $2000 allowable limit for IRAs, and can go as high as 25% of total income. The funds accumulate in an investment portfolio, and become available after age 59½.

Assume $1000 is to be set aside in a tax-deferred retirement account. What is the advantage of tax deferral over after-tax investing? Table 6.10 compares the growth of a tax-deferred and after-tax investment over 30 years. The after-tax investment has a lower starting principal, and tax must be paid on the interest earned each year. The tax-deferred investment is not taxed until the money is withdrawn. The table assumes that the portfolio earns an effective rate of 10% per year in both cases. It also assumes a 34% tax bracket, which is the national average.

Table 6.10. A Comparison of Tax-Deferred and After-Tax Investments

Tax-deferred investment:

money available − tax = total invested
$1000 − 0   = $1000

Year	Value	Interest earned	After-tax value if withdrawn
5	$1,611	611	$ 1,063
10	2,594	1,594	1,712
15	4,177	3,177	2,757
20	6,727	5,727	4,440
25	10,835	9,835	7,151
30	17,449	16,449	11,516

After-tax investment:

money available − tax = total invested
$1000 − 340 = $660

Year	Value	Interest earned	After-tax value if withdrawn
5	$1,063	403	$ 926
10	1,712	1,051	1,354
15	2,757	2,097	2,044
20	4,440	3,780	3,155
25	7,151	6,491	4,944
30	11,517	10,857	7,826

This example is oversimplified, but it does show that there is a considerable cumulative advantage to investing pretax dollars. The 34% figure used to calculate the tax on the interest earned from the after-tax investment is excessive. But even at a lower rate, the tax-deferred investment earns enough to offset even a 10% early-withdrawal penalty.

INDIVIDUAL RETIREMENT ACCOUNTS

There are few restrictions on who may open and contribute to an IRA. Furthermore, the tax-deductibility of the yearly contribution can be complicated, partly because the tax laws covering IRAs have undergone several changes since they were instituted. In particular, in 1986 the qualifications for the full IRA deduction were tightened, and many people pulled back on their investments in these programs. Those with high incomes were hit hardest by the change. Today the deductibility of these pension plans depends on adjusted gross income, marital status, and whether the contributor already participates in a sponsored retirement plan. What follows is a basic outline of eligibility requirements for IRAs. (For more details, see IRS Publication 590.)

1. Allowable Contributions Any individual may contribute to an IRA account up to $2000 per year or 100% of earned income, whichever is less. If husband and wife both work, each may contribute up to $2000 per year. A couple with one wage earner, when filing jointly, may contribute up to $2250 per year (the extra $250 would go into a separate IRA).

2. Allowable Deductions The full IRA tax deduction of $2000 is allowed if the contributor is not covered by an employer-sponsored pension plan. A contributor who *does* belong to a pension plan may still deduct the full $2000 IRA contribution if:

    i. adjusted gross income is less than $40,000 (on a joint return), or
    ii. adjusted gross income is less than $25,000 (when filing single).

A partial deduction is allowed if:

    i. adjusted gross income is greater than $40,000 but less than $50,000 (on a joint return), or
    ii. adjusted gross income is greater than $25,000 but less than $35,000 (when filing single).

No deduction is allowed above these limits.

3. Deadlines and Penalties Money can be contributed to an IRA at any time up until April 15 and still be deducted for the preceding tax year. In addition, a first IRA may be opened by April 15 in order to establish a deduction for the previous calendar year.

Tax-deferred money that is withdrawn before age 59½ is subject to a 10% penalty levied by the IRS, in addition to the tax that must be paid. However some penalty-free withdrawals are possible if they are part of a regular withdrawal schedule in which withdrawn amounts are based on life expectancy and the expected rate of return of the IRA. Under such a plan, withdrawals must be made for at least five years (or until age 59½), after which withdrawals can be made freely with no penalty.

To calculate the payment schedule for early withdrawals, divide total IRA savings by life expectancy and estimate the expected rate of return. (There is a maximum rate allowed by the IRS.)

4. Taking Money out of an IRA After age 70½ no money may be contributed to an IRA. Furthermore, money must be withdrawn yearly in an amount proportional to the contributor's remaining life expectancy. The IRS mandates this so that taxes will be paid on the tax-deferred income. If the contributor does not withdraw the minimum amount, the IRS levies a 50% penalty tax on the difference.

ROTH IRAS

Introduced in 1998, the Roth IRA presents an interesting twist on the traditional IRA. The annual contribution to a Roth IRA (up to $2000) is *not* tax-deductible. However, earnings as well as contributions are tax-free upon withdrawal, provided at least one of the following conditions is met:

the contribution has been invested for at least five years.
the investor has attained the age of 59½.
the investor has incurred a disability.
the investor has purchased a first home.
the investor dies.

In the absence of at least one of these circumstances, any distribution of Roth IRA earnings is included in the investor's gross income for the year of withdrawal. Furthermore, there are penalties for withdrawals made before the age of 59½. Aside from the tax advantage it offers, a Roth IRA is less restrictive in its age requirements than a traditional IRA. Specifically, Roth withdrawals are not mandated for those who attain the age of 70½, nor are contributions denied to anyone beyond that age.

The choice between a Roth and a traditional IRA rests upon more factors than can be briefly explained here. A comparative analysis, which any investment analyst should be able to produce, is essential for any interested investor.

SEP-IRAS

A *simplified employee pension plan* (SEP-IRA) is an option available to those who earn a substantial part of their income through self-employment. In an SEP-IRA, an individual can make a tax-deductible contribution to an investment program of as much as 15% of self-employment income (to a maximum of $22,500). This option assumes that the individual is not part of any other retirement program.

The advantage of an SEP-IRA is clearly in the amount of allowable contributions. In an ordinary IRA $2000 is the annual limit; in an SEP-IRA the limit is $22,500.

KEOGH PLANS

Keogh plans were designed for employees in small businesses such as accounting, law, architecture, and engineering firms. A Keogh plan requires a lot of effort to set up, but it allows higher yearly contributions than an SEP.

There are three basic types of Keogh plans: profit sharing, money purchase, and paired plan.

> *Profit Sharing:* Of all Keogh plans, this option has the lowest allowed annual contribution, just 15% of earned income (up to no more than $30,000 per participant). But profit sharing is the most flexible of the Keogh plans because it allows the contribution to vary from year to year, from as little as nothing to as high as the dollar or percent limit.
>
> *Money Purchase:* This plan allows a yearly contribution of up to 25% of earned income, but not more than $30,000. However, it requires the contributor to specify a fixed annual contribution, and this contribution must be met every year until the contributor reaches age 59½.
>
> *Paired Plan:* The paired plan stipulates that the contributor invest the maximum allowable amount each year. This may be determined by a percent of salary (25%) or by the $30,000 allowable limit, whichever is lower.

Banks and financial services offer ready-made SEP and Keogh plans featuring many different types of investment portfolios. Like IRAs, these plans defer taxes on a portion of income, allowing the money to earn interest until the contributor reaches age 59½, after which he or she may begin to withdraw the money without penalty. Income tax must be paid on the withdrawal as though it were income in the tax year of the withdrawal.

401(K) AND 403(B)

The names 401(k) and 403(b) come from item numbers of the 1978 tax code that created these retirement plans. 401(k) is a tax-deferred income investment plan available to corporate employees. 403(b) is designed for employees of public schools and colleges, and other nonprofit organizations.

In these plans, the employee can make pretax contributions of up to 15% of salary (with an upper limit that is adjusted every year) to an investment fund set up and managed by the employer (with input from the employees). Employers usually match some percentage of the employee contribution. The pool of 401(k) or 403(b) money can be invested in several ways—stock or bond funds, money market funds, or GICs (guaranteed income contracts). The tax advantage is similar to that of IRA and Keogh plans. Taxes on income are deferred until the money is withdrawn after retirement. Not only will taxes be paid after peak earning years (and thus at lower rate), but the deferring of taxes allows the fund to earn interest during the years until retirement.

Like all payroll-deduction plans, 401(k) plans are examples of dollar cost averaging; see page 311.

Investment Strategies

Designing an investment portfolio involves many choices and some degree of uncertainty. The inflation rate, and thus the value of a dollar in the future, looms as the primary consideration in deciding how much saving is enough and what return on investment is adequate to meet future expenses. Because of this uncertainty, exact mathematical formulas are not necessary. Instead, an investor needs some guidelines that establish minimum levels of investment and earnings in order to meet future goals. Here are three such guidelines.

1. The Effects of Inflation on Future Costs This formula shows how much something costing P dollars today will cost several years from now. It is nothing more than a compound interest calculation that shows the rise in price due to inflation:
$$F = P \times (1 + i)^n$$
where P is the cost today, F is the cost n years from now, and i is the estimated annual inflation rate.

* A vacation house is priced at $250,000. Assuming a 3% annual rate of inflation, a comparable house 10 years from now would cost
$$250,000 \times (1 + .03)^{10} = \$335,979 \approx \$336,000$$

2. Estimating Return on Investment That Will Maintain Purchasing Power

Because the interest earned on investments will be reduced by taxes, the rate of return should exceed the rate of inflation in order to maintain purchasing power. The required before-tax rate of return (r) should be

$$r = \frac{i}{1 - t}$$

where i is the annual inflation rate (in decimal form), and t is the income-tax bracket (also a percent in decimal form).

♦ An investor in the 34% tax bracket estimates a 3% rate of inflation. To keep pace with inflation, she must have a before-tax annual yield of

$$r = \frac{0.03}{1 - 0.34} = 0.04545 \approx 4.5\%$$

Table 6.11 shows how after-tax yield depends upon one's tax bracket. For example, someone in the 34% tax bracket who wishes to maintain an after-tax annual return on investment of 9% must have a portfolio that earns 13.64%.

Table 6.11. Annual Return on Investments Required to Offset Taxes

Tax bracket and corresponding annual pretax yields

After-tax yield	28%	30%	31%	32%	34%	36%	37%	39%
5½%	7.64	7.86	7.97	8.09	8.33	8.59	8.73	9.02
6%	8.33	8.57	8.70	8.82	9.09	9.38	9.52	9.84
6½%	9.03	9.29	9.42	9.56	9.85	10.16	10.32	10.66
7%	9.72	10.00	10.14	10.29	10.61	10.94	11.11	11.48
7½%	10.42	10.71	10.87	11.03	11.36	11.72	11.9	12.30
8%	11.11	11.42	11.59	11.76	12.12	12.50	12.70	13.11
8½%	11.81	12.14	12.32	12.50	12.88	13.28	13.49	13.93
9%	12.50	12.86	13.04	13.24	13.64	14.06	14.29	14.75
9½%	13.19	13.57	13.77	13.97	14.39	14.84	15.08	15.57
10%	13.89	14.29	14.49	14.71	15.15	15.63	15.87	16.39
10½%	14.58	15.00	15.22	15.44	15.91	16.41	16.67	17.21
11%	15.28	15.71	15.94	16.18	16.67	17.19	17.47	18.02

3. Necessary Rate of Return When Funds Are Withdrawn

Assuming an investment portfolio is providing regular payments to meet living expenses, the before-tax rate of return calculated in paragraph 2 above would have to be higher to offset both the taxes paid and the amount withdrawn each year.

In this formula (a refinement of the formula in paragraph 2), the annual withdrawal from the portfolio, given as a percent, is w. The rate of return (r) should be

$$r = \frac{i + w}{1 - t}$$

♦ Assume that 4% of a portfolio is withdrawn each year and the annual rate of inflation is 3% ($i = 0.03$). The target rate of return for an investor in the 34% tax bracket ($t = 0.34$) would be

$$r = \frac{.03 + .04}{1 - .34} = 0.10606 \approx 10.6\%$$

DOLLAR COST AVERAGING

Some investors follow a policy of investing a fixed sum at regular intervals. (One possibility is a monthly lump-sum investment through a payroll deduction.) The practice of purchasing a fixed-dollar value of shares at regular intervals is called *dollar cost averaging,* and it lends itself particularly well to mutual-fund purchases because the investor may purchase any dollar amount instead of multiples of the price per share. Dollar cost averaging may also be applied to the purchase of stocks or bonds.

Variations on dollar cost averaging sound much like betting systems on casino games, with an important difference: your expected gain is much better. Still, investments are a form of gambling, and these systems, like the variations on the bettor's martingale (see page 228), depend on *how much* to invest as opposed to *when* to invest. These are the basic averaging systems:

i. *Straight dollar cost averaging:* Choose an investment; invest a fixed sum every month.

ii. *Accelerated dollar cost averaging:* Choose an investment; set an amount as a monthly investment goal; vary the amount each month depending on the purchase price. If the price is high, buy less; if low, buy more according to this formula:

$$\text{monthly investment} = \text{monthly goal} \times \left(\frac{\text{average price of stock}}{\text{current price}} \right)$$

iii. *Value averaging:* Choose a no-load or low-load investment and invest enough each month to increase your total investment by a fixed amount.

♦ If you have $225 a month to invest, straight dollar cost averaging is the simplest method and could possibly be arranged through a payroll deduction. By the accelerated method, you would invest in a stock and keep track of the average price per share you have been paying. If, for example, the average price is $20 per share and this month it is down $2 to $18 per share, you would purchase more according to this formula:

$$\text{monthly investment} = \$225 \times \left(\frac{20}{18}\right) = \$250$$

When the price is up, the ratio of average price to current price decreases your monthly allotment.

Finally, in value averaging, you would calculate how much you have to invest in order to increase your account by $225 each month. If the investment posted big gains during a month, your contribution would be less than $225.

Basic Bookkeeping

The object of bookkeeping is to record and summarize all business transactions that take place within a given time period. The same methods apply whether the business is a single household or a multinational conglomerate. A journal of some type is used to record every transaction involving money or possessions with monetary value. The journal entries are periodically transferred to an account ledger (a process called *posting*), where they are grouped by type. These entries are ultimately transferred to a summary statement—a monthly, quarterly, or annual report in the form of a *balance sheet* and an *income statement*.

ASSETS AND LIABILITIES

An *asset* is some property or resource whose use or value is available to the owner. A *liability* is an obligation or debt representing someone else's advantage. A *balance sheet* is a listing of assets and liabilities, designed to show how possessions and debts are balanced against each other. The difference between assets and liabilities is what a company or individual actually owns—that is, its *net worth*.

Assets take the form of cash, checks, and other cashable notes, accounts receivable, supplies, inventory, prepaid items, land, buildings, vehicles, equipment, and so on. Essentially, an asset is anything that has a cash value or could be converted to cash if necessary.

Liabilities are broken into two categories: short-term and long-term liabilities. Both categories represent types of debt or bills that will have to be paid. Short-term debts include monthly utility bills, payroll, and any short-term loans that are coming due. Long-term debts include mortgages as well as money that has been raised from (and is therefore owed to) stock and bond holders by a stock-issuing corporation.

Equity is defined as what would be left over if a company or individual paid off all debts.

THE BALANCE SHEET

The purpose of a balance sheet is to show the breakdown of assets, liabilities, and equity. These quantities are related by a simple equation:

$$\text{assets} = \text{liabilities} + \text{owner's equity}$$

The balance-sheet equation may apply to an entire business, to a household budget, or to a specific investment such as a house. For example, a $120,000 house purchased with a 20% down payment is a capital asset. The value of the asset (the house) can be broken down into the liability (the mortgage) and the owner's equity (the down payment).

$$\$120,000 = \$96,000 + \$24,000$$

$$\text{asset} = \text{mortgage} + \text{equity}$$

As the mortgage is paid off, this equation changes. The owner gradually accumulates more equity as the principal is reduced. This process is called *amortization* (see page 275).

Table 6.12 shows a typical balance sheet that follows standard formatting practices. A single underline is used for the last figure in a column of added numbers or under any number that is subtracted. Thus a single underline is used for every subtotal or last figure in a calcula-

Table 6.12. A Typical Balance Sheet

The Baxter Company
Balance Sheet
Dec 31, 19__

Current Assets			Current Liabilities		
Cash	$30,000		Accounts payable	$7,000	
Inventory	15,000		Wages payable	10,000	
Prepaid insurance	3,000		Note payable	5,000	
Prepaid services	1,000				
Total current assets		$49,000	Total current liabilities		$22,000
Long-term Assets			**Long-term Liabilities**		
Computers	$10,000		Mortgage payable	$35,000	
Office furniture	4,000		Total long-term liabilities		$35,000
Machinery	3,000				
Total long-term assets		$17,000	Total liabilities		$57,000
			Owner's equity		9,000
Total Assets		$66,000	**Total Liabilities and Equity**		$66,000

tion. A double underline is used for final totals. A dollar sign is placed before any figure at the top of a column containing figures that represent sums of money, and also placed before any dollar figure that is the result of a computation involving figures above it—that is, before any total or subtotal.

THE INCOME STATEMENT

Net income is the difference between the amount of money taken in and the amount of money paid out in a given period of time. This is summed up by the statement "Profit equals revenue minus cost." The purpose of an income statement is to record revenues and costs in order to state the *bottom line*—the profit or loss for that period—which is literally the last line of the statement.

Income statements cover specific periods of time: a month, a quarter, six months, or a year. For businesses, the annual income statement corresponds to the calendar year January 1–December 31, which is also the tax year.

A typical income statement is shown in Table 6.13. It itemizes revenues and costs relevant to a particular company. The variety of costs and revenues is not limited to those listed here. They can vary widely in type, but the essential structure and purpose of an income statement remain consistent, whether it applies to a household, a small business, or a large corporation.

Table 6.13. A Typical Income Statement

Acme Appliance Company
Income Statement
for the year ending Dec 31, 19__

Revenues		
Sales	$150,000	
Service	75,000	
Total revenues		$225,000
Expenses		
Wages	$110,000	
Rent	60,000	
Insurance	10,000	
Taxes	15,000	
Advertising	5,000	
Utilities	4,000	
Total Expenses		$204,000
Net Income		$21,000

DOUBLE-ENTRY BOOKKEEPING

In traditional bookkeeping, every business transaction is recorded in a journal, which is a chronological record of transactions. Journal entries correspond to receipts such as invoices, shipping receipts, payments stubs, and sales slips. In a double-entry system, each transaction is recorded twice, once as a credit to one account and once as a debit to another account.

From the journal, debits and credits are transferred to a ledger, in which they are grouped by type (instead of date); that is, cash transactions, individual customer accounts, and other specific classes of records are isolated so that, for example, an individual customer's account can be examined.

The purpose of double-entry bookkeeping—which is standard even when the journal and ledger are computer databases rather than bound books—is to maintain a reliable record of transactions. Because each transaction is recorded twice, once as a debit and once as a credit, the sum of the debits must match the sum of the credits. This is known as *balancing the books*.

A traditional rule of thumb for balancing the books is that *all transposition errors create discrepancies that are multiples of 9*. If two digits are transposed (reversed) when entered into the journal, cash register, or checkbook, the mistake will be noticed when debits do not match credits. If the discrepancy is divisible by 9 (recall that a number is divisible by 9 if the sum of its digits is divisible by 9), it is probably a transposition error. This is a rule long observed by all cashiers when they cash out.

DEPRECIATION

Because assets in the form of durable goods suffer wear and tear, they do not retain their value over time. With use, they depreciate in value. This is true not only in fact (when trying to resell, for example) but for accounting purposes. The assets listed on a balance sheet will change in value as they wear out.

A car, for example, has a *book value* that is largely determined by its age. Most of the depreciation in the value of a car or any other heavily used piece of equipment occurs in the first few years of use. It is not uncommon, in fact, for cars, computers, office equipment, and machinery to have no more than five or six years of useful life. Consequently, accounting methods have been developed to allow owners to accurately state the true value of assets. All of the methods work from three figures: the *original cost* (or initial value) of the asset, its expected *useful lifetime*, and its *salvage value*. How the initial value of the asset is depreciated depends on how it is used. In accounting, there are four principal methods that can be followed: *straight-line depreciation, sum-of-the-years-digits, double-declining balance,* and *units of production.*

The first of these is the simplest because it involves depreciating the same amount each year. The second and third methods depreciate the greatest amount in the first year, with lesser percentages in succeeding years. These are called *accelerated depreciation* methods. The last method is one that depreciates based on the amount of usage rather than the time of ownership.

For tax purposes, the IRS outlines several acceptable methods of straight-line and accelerated depreciation. These are described in IRS Form 4562.

1. Straight-Line Depreciation The straight-line method depreciates equal amounts for each year of useful life. The value that remains after the useful life has elapsed is called the *salvage value.*

$$\text{yearly depreciation} = \frac{\text{initial cost} - \text{salvage value}}{\text{estimated life}}$$

◆ A company car cost \$16,500 new. It has an estimated useful life of six years, after which it will be sold for \$4500. Using straight-line depreciation, the yearly depreciation is

$$\frac{\$16,500 - 4500}{6} = \frac{12,000}{6} = \$2000$$

If the company needs to keep track of *monthly depreciation,* it would divide this figure by 12.

$$\text{monthly depreciation} = \frac{\text{yearly depreciation}}{12}$$

2. Sum-of-the-Years-Digits Method Unlike straight-line depreciation, which depreciates the same fraction of the useful value of an asset each year, the sum-of-the-years-digits method employs a sequence of decreasing fractions. These are based on the expected lifetime of the asset. For example, if an asset is allotted a useful lifetime of five years, the method calls for summing $5 + 4 + 3 + 2 + 1$ to get 15. This sum of the digits is then used as the denominator of a sequence of five fractions—$5/15$, $4/15$, $3/15$, $2/15$, and $1/15$—that represent the portion of the asset's value depreciated each year.

♦ Assume the car in the example above is depreciated over six years using the sum-of-the-years-digits method.

$$6 + 5 + 4 + 3 + 2 + 1 = 21$$

Thus the fractions are: $\frac{6}{21}$, $\frac{5}{21}$, $\frac{4}{21}$, $\frac{3}{21}$, $\frac{2}{21}$, and $\frac{1}{21}$. The usable value is

initial cost − salvage value = $16,500 − 4500 = $12,000

1st-year depreciation: $\frac{6}{21}$ × $12,000 = $3428.57

2nd-year depreciation: $\frac{5}{21}$ × $12,000 = 2857.14

3rd-year depreciation: $\frac{4}{21}$ × $12,000 = 2285.71

4th-year depreciation: $\frac{3}{21}$ × $12,000 = 1714.29

5th-year depreciation: $\frac{2}{21}$ × $12,000 = 1142.86

6th-year depreciation: $\frac{1}{21}$ × $12,000 = 571.43

Total depreciation $12,000.00

3. Double-Declining Balance Method This method begins by calculating the percent used in straight-line depreciation and doubling it. Each year's depreciation is the same percent of the *remaining* value of the asset. Salvage value is ignored until the last year's depreciation has been found.

♦ If a $40,000 machine is depreciated over five years, the straight-line depreciation over one year (ignoring salvage value) would be $\frac{1}{5}$ (or 20%) of the value each year. In the double-declining balance method, this percentage is doubled and used to find the depreciation in the first year. In the second year, depreciation is the same percentage of the remaining value, and so on.

Year	Initial value	Depreciation (40% of initial value)	
1	$40,000	0.4 × 40,000 =	$16,000.00
2	24,000	0.4 × 24,000 =	9,600.00
3	14,400	0.4 × 14,400 =	5,760.00
4	8,640	0.4 × 8640 =	3,456.00
5	5,184	0.4 × 5184 =	2,073.60
		Total depreciation	$36,889.60

salvage value = initial value − total depreciation
= $40,000 − 36,889.60
= $3110.40

4. Units-of-Production Method This method is based on usage rather than on the time an asset has been owned. It is primarily used for machinery that has an expected operational lifetime that can be measured in units of service. A car, for example, might be depreciated based on the number of miles it has been driven rather than how old it is. A stamping machine might be rated for 50,000 stampings before it wears out. The units-of-production method depreciates according to the percent of useful service that has been used up. The method consists of three steps:

   i. Find the asset's usable value by subtracting salvage value from original cost.
  ii. Divide the usable value by the number of units of useful service (in miles, hours of operation, individual uses).
 iii. Multiply the initial value by the ratio of units used to total units.

♦ A robot welder with an initial value of $240,000 is built to perform 400,000 welds before it wears out. If it performs 100,000 welds in its first year of use, it has used up $100,000/400,000$, or one-quarter, of its value. Thus, by the units-of-production method, a fourth of $240,000, or $60,000, can be deducted in the first year.

PAYROLL

A company's payroll breaks down into two components: salary and wages. *Salary* is compensation paid to employees who are hired on contract for specified periods of time. Some employees may be hired for only a few weeks or a month, but most salaried employees have renewable contracts that provide a measure of job security.

Wages constitute the money paid to hourly workers—usually long-term non-management employees or short-term outside contractors. These workers must keep a record of their hours worked, either by punching a time clock or by submitting accurate accounts. Hourly compensation is set at a predetermined level. Thus a wage earner's weekly salary is found by multiplying the number of hours worked by the hourly pay rate. The traditional work week in the United States is a 40-hour week. These hours—the normal working hours from 9:00 to 5:00 Monday through Friday—are referred to as *straight time.* An employee who works more than 40 hours in a given week may be eligible for a higher hourly rate on the extra hours. The extra hours and the extra money are referred to as *overtime.*

The hourly wage for overtime is given as a multiple of the regular hourly wage. It can vary with the type of overtime. The following overtime schedule is typical:

Time-and-a-half: Extra weekday work beyond 40 hours
Double time: Sundays and holidays
Variable up to triple time: Evening or emergency shifts

♦ An $8.40/hour wage earner works 15 hours of overtime one week. The overtime pay is time-and-a-half, or

$$1\frac{1}{2} \times \$8.40 = \$12.60/\text{hour}$$

Her gross pay for the week will be:

Regular pay: 40 × $8.40	=	$336.00
Overtime pay: 15 × $12.60	=	189.00
Gross pay:		$525.00

PAYROLL DEDUCTIONS

By law, all businesses must deduct state and federal taxes from paychecks. These withholdings are paid directly to the government. Other deductions may also be made for the employee's contribution to a health care program, unemployment compensation, Social Security, Medicare, elective tax-deferred retirement program, day care, insurance, or additional prepaid taxes.

The pay statement lists the total pay before deductions. This is the *gross income.* The amount of pay after deductions is the *net income.* Although the deductions are itemized on a pay statement, their designations can be cryptic. *FICA,* for example, refers to Social Security payments. By law, the employer must match the employee's Social Security contribution. Although the exact percentage can vary year by year, it has recently been 7%. Thus the employer must contribute 7% of the employee's gross pay in addition to the 7% employee contribution. The total contribution is 14%.

Workers' compensation is a fund maintained by contributions from employers. The contribution for any employer is usually a percentage of total payroll, and may be paid to the state or to a private insurance provider, depending on local laws.

An employer must file a quarterly report of federal tax withheld on paychecks. At the end of the fiscal year, the employer must also provide each employee with a W-2 form, which states the total wages or salary earned, the total deductions for federal, state, and local taxes, and the total for other deductions that might be tax-deferred.

Individual Income Tax

Individual income tax is calculated according to one basic formula:

taxable income = adjusted gross income − deductions
− exemptions − credits

The tax-return form is divided into sections in which totals for each of these categories are computed. When the taxable income is found, the tax payment can be found in a *tax table*. The formula varies in percent amounts with different categories of taxpayer: income up to the limit of the lowest tax bracket is calculated at the lowest tax rate; income that falls into the next tax bracket is calculated at the next tax rate; and so on.

Table 6.14 and the example that follows it show how the tax in the tax table is derived. Note that someone in the 31% tax bracket does not pay 31% on all taxable income. The rate of 31% applies only to the amount of taxable income that exceeds $99,600 (on a joint tax return). The 28% rate applies to taxable income that falls between $41,200 and $99,600 and the lowest rate—15%—applies to income less than $41,200. Thus even millionaires pay only 15% on the first $40,000 or so ($20,000 in the case of single millionaires) of their income.

Table 6.14. Tax brackets (1997)

Single	Joint	Rate
$0–24,650	$0–41,200	15%
$24,651–59,750	$41,201– 99,600	28%
$59,751–124,650	$99,601–151,750	31%
$124,651–271,050	$151,751–271,050	36%
$271,051 and up	$271,051 and up	39.6%

• A joint tax return that shows a taxable income on $110,000 would owe a tax that is calculated as follows:

$$
\begin{aligned}
\$41,200 \times 0.15 &= \$6,180.00 \\
(\$99,600 - 41,200) \times 0.28 &= 16,352.00 \\
(\$110,000 - 99,600) \times 0.31 &= \underline{3,224.00} \\
\text{Total tax} &\qquad \$25,756.00
\end{aligned}
$$

(This calculation is spelled out in more detail in the instruction book for IRS form 1040.)

THE 1040 FORM

A taxpayer may file a return using a 1040, 1040A, or 1040EZ form, or electronically by e-File or Telefile. Each of the forms has the same basic format, the 1040 being the longest and most detailed. They are all designed to arrive at a figure for taxable income, which is based on adjusted gross income minus certain allowable deductions, exemptions, and credits. These terms are briefly explained below.

1. Adjusted gross income The first part of a tax return covers all sources of income. When income that is not subject to tax (welfare payments, interest on municipal bonds, etc.) is deducted from the total income, the result is called adjusted gross income.

Adjusted gross income is the total of all wages, salary, unemployment compensation, tips, gratuities, interest and dividend earnings, annuities, royalties, rents, Social Security income, and some other sources of income. It excludes income that is not subject to tax. Thus payments such as public assistance (welfare) and interest on tax-exempt bonds are subtracted from total income to arrive at adjusted gross income. Also subtracted are any amounts on which tax will be paid at a future date or by another party. This includes alimony payments and penalties on early withdrawal of savings, as well as any payments to tax-deferred retirements plans (IRAs, Keogh plans, etc.).

2. Deductions Adjusted gross income is further reduced by certain deductions that are sanctioned by the tax laws. Everyone is entitled to take a *standard deduction.* In a recent year, the standard deductions were:

single: $4150
head of household: $6050
married filing jointly: $6900
married filing separately: $3450
over 65 or blind: additional $1000 (single)
additional $800 (married)

Individual items that can be deducted from taxable income include state and local property taxes, mortgage interest payments, charitable contributions, moving expenses related to job relocation, medical expenses (when in excess of 7.5% of adjusted gross income), and casualty losses.

When the sum of these deductions exceeds the standard deduction, the taxpayer may choose to *itemize.* This involves adding up all allowable deductions and subtracting the sum from the adjusted gross income.

3. Exemptions In addition to deductions, every taxpayer is allowed to take a personal *exemption,* and an exemption of the same amount for a spouse and for each dependent not claimed on another taxpayer's return.

4. Tax Credits Taxable income can also be reduced if the taxpayer is eligible for a *tax credit*. This is a deduction available to the poor, the elderly, and those who bear expenses related to the care of a dependent. Schedules for each type of tax credit can be found in the tax booklet.

Φ ———————————————————————————————— Φ

TAX FREEDOM DAY

Tax Freedom Day is the day of the year on which the average taxpayer has earned enough to cover his or her annual taxes (state, local, and federal). It is an excellent example of the many misleading statistical "facts" that circulate widely.

Tax Freedom Day is a calendar method of expressing the percentage of gross income that goes to pay taxes. For the average American, taxes account for about 35% of income. (For a brief discussion of "average," see page 241.) In 1996 Tax Freedom Day fell on May 7, or 128 days into a 366-day year (35% of 366 is 128). The same idea can be applied to an eight-hour workday. By this reckoning, the average American in 1996 did not begin working for himself or herself until 2 hours and 48 minutes into each eight-hour workday (35% of 8 hours is 2.8 hours); everything up to that point was for the government.

The problem with this statistic is that it encompasses such a wide range of incomes as to be almost meaningless for any individual. Also, it suggests that all of this money constitutes income tax, whereas in fact it includes sales tax, excise tax, and property tax as well. And almost a third of it consists of payments to social insurance programs such as Social Security, Medicare, and Medicaid, money that comes back to "average Americans" in one way or another. Thus to say that money earned prior to Tax Freedom Day is entirely "for the government" overlooks the extent to which the government *is* the people.

Φ ———————————————————————————————— Φ

Appendix: Tables and Formulas

Mathematical Symbols

ROMAN NUMERALS

In the system of Roman numerals, repeated symbols indicate added values. (Thus XXX is three 10s, which equals 30.) A symbol preceded by a symbol of lower value is reduced by the lower value. (Thus CM is 100 less than 1000, which equals 900.) A symbol followed by a symbol of lesser value is increased by the lesser value. (Thus MCM is 1000 followed by 900, which is 1900.) A symbol with an overline is multiplied in value by 1000. (Thus $\overline{X}$ is 10×1000, which equals 10,000.)

1	I
2	II
3	III
4	IV
5	V
6	VI
7	VII
8	VIII
9	IX
10	X
11	XI
12	XII
13	XIII

14	XIV
15	XV
16	XVI
17	XVII
18	XVIII
19	XIX
20	XX
25	XXV
30	XXX
40	XL
50	L
60	LX
70	LXX
80	LXXX
90	XC
100	C
150	CL
200	CC
300	CCC
400	CD
500	D
600	DC
700	DCC
800	DCCC
900	CM
1000	M
1500	MD
1600	MDC
1650	MDCL
1700	MDCC
1750	MDCCL
1800	MDCCC
1850	MDCCCL
1860	MDCCCLX
1870	MDCCCLXX
1880	MDCCCLXXX
1890	MDCCCXC
1900	MCM or MDCCCC
2000	MM
3000	MMM
4000	MMMM or $M\overline{V}$
5000	$\overline{V}$
10,000	$\overline{X}$
50,000	$\overline{L}$
100,000	$\overline{C}$
1,000,000	$\overline{M}$

Thus, MCMXLVII is M + CM + XL + V + II = 1000 + 900 + 40 + 5 + 2 = 1947.

GREEK ALPHABET

Greek letters are commonly used as symbols for mathematical constants, angle measures, and variables.

Symbol	Name	English equivalent
A, α	Alpha	a
B, β	Beta	b
Γ, γ	Gamma	g
Δ, δ	Delta	d
E, ε	Epsilon	e
Z, ζ	Zeta	z
H, η	Eta	ē
Θ, θ, ϑ	Theta	th
I, ι	Iota	i
K, κ	Kappa	k
Λ, λ	Lambda	l
M, μ	Mu	m
N, ν	Nu	n
Ξ, ξ	Xi	x
O, o	Omicron	o
Π, π	Pi	p
P, ρ	Rho	r
Σ, σ	Sigma	s
T, τ	Tau	t
Υ, υ	Upsilon	u
Φ, φ, φ	Phi	ph
X, χ	Chi	ch
Ψ, ψ	Psi	ps
Ω, ω	Omega	ō

MATH SYMBOLS

+ plus or positive
− minus or negative
×, ·, * times; multiplied by
÷, / divided by
= is equal to
% percent
> is greater than
< is less than

≥   is greater than or equal to
≤   is less than or equal to
≠   is not equal to
≡   is equivalent to; is defined as; is identical with
≅   is approximately equal to; is congruent to
≈   is approximately equal to
±   plus or minus
√   square root of
∏   product of
∠   angle
∪   union (of sets)
∩   intersection (of sets)
∅   empty set or null set
∝   is directly proportional to
∞   infinity
π   pi (approximately 3.14159)
i   the square root of -1
∴   therefore (used in proofs)

Units of Measurement

UNITS OF TIME
Time Divisions

1 minute = 60 seconds
1 hour = 60 minutes
1 day = 24 hours
1 week = 7 days
1 fortnight = 2 weeks
1 month = 28–31 days
1 year = 12 months; 365 days
1 decade = 10 years
1 century = 100 years
1 millennium = 1000 years

Regular time intervals

daily	once a day
weekly	once a week
semiweekly	twice a week
biweekly	once every two weeks *or* twice a week

triweekly	once every three weeks
monthly	once a month
bimonthly	once every two months *or* twice a month
quarterly	once every three months (usually beginning January 1)
semiannual	once every six months
annual	once a year
biennial	once every two years; lasting two years
biannual	once every six months
decennial	once every 10 years; lasting 10 years; covering 10 years

Observances

centennial	100-year anniversary
bicentennial	200-year anniversary
tercentennial	300-year anniversary
millennial	relating to a 1000-year anniversary
sesquicentennial	150-year anniversary

Divisions of the solar year (Northern Hemisphere)

vernal equinox	the first day of spring, on or about March 21
autumnal equinox	the first day of autumn, on or about September 21
summer solstice	the first day of summer and the day of the sunlight's longest duration, on or about June 21
winter solstice	the first day of winter and the day of the sunlight's shortest duration, on or about December 21

UNITS OF LENGTH

1 palm = 3 inches
1 hand = 4 inches (used for heights of horses)
1 span = 9 inches (based on span from thumb to little finger of outstretched hand)
1 foot = 12 inches
1 cubit = 18 inches (based on length of a hand and forearm)
1 yard = 3 feet
1 fathom = 6 feet = 2 yards
1 rod = 5½ yards = 16½ feet
1 furlong = 40 rods = 660 feet (based on the length of a furrow)

1 mile = 8 furlongs = 1760 yards = 5280 feet
1 league ≈ 3 miles

UNITS OF AREA

1 square foot = 144 square inches
1 square yard = 9 square feet
1 square rod = 30¼ square yards
1 acre = 160 square rods = 4840 square yards = 43,560 square feet (based on the area a yoke of oxen could plow in one day)
1 square mile = 640 acres

UNITS USED BY SURVEYORS

1 link = 7.92 inches
1 rod = 25 links = 16.5 feet
1 chain = 4 rods = 100 links = 66 feet
1 furlong = 10 chains = 220 yards = ⅛ mile
1 mile = 80 chains
1 square link = 62.73 square inches
1 square rod = 625 square links
1 square chain = 16 square rods = 1000 square links
1 acre = 160 square rods = 10 square chains
1 square furlong = 10 acres
1 square mile (or section) = 640 acres = 64 square furlongs
1 township = 36 square miles (6 miles × 6 miles of area)

UNITS OF VOLUME

1 cubic foot = 1728 cubic inches (12 inches × 12 inches × 12 inches)
1 cubic yard = 27 cubic feet (3 inches × 3 inches × 3 inches)
1 liter = 1000 cubic centimeters (10 centimeters × 10 centimeters × 10 centimeters)
1 board foot = 144 cubic inches (12 inches × 12 inches × 1 inch)
1 cord foot = 16 cubic feet (4 feet × 4 feet × 1 foot)
1 cord = 8 cord feet = 128 cubic feet (4 feet × 4 feet × 8 feet)

UNITS OF DRY MEASURE

1 quart = 2 pints
1 peck = 8 quarts = 16 pints
1 bushel = 4 pecks

UNITS OF (AVOIRDUPOIS) WEIGHT

1 dram = 27.344 grains
1 ounce = 16 drams
1 pound = 16 ounces
1 stone = 14 pounds
1 quarter = 28 pounds = 2 stone
1 short hundredweight = 100 pounds
1 long hundredweight = 112 pounds = 4 quarters
1 short ton = 2000 pounds
1 long ton = 2240 pounds

UNITS OF LIQUID MEASURE (U.S. SYSTEM)

1 tablespoon = 3 teaspoons = 0.5 fluid ounce
1 gill = 4 fluid ounces
1 cup = 8 fluid ounces
1 pint = 2 cups = 16 fluid ounces
1 quart = 2 pints = 4 cups = 32 fluid ounces
1 gallon = 4 quarts = 16 cups = 128 fluid ounces
1 barrel (ordinary liquid) = 31.5 gallons
1 barrel (oil) = 42 gallons
1 barrel (beer) = 43.25 gallons
1 hogshead = 63 gallons
1 pipe = 2 hogsheads
1 tun = 2 pipes

NAUTICAL MEASURES

1 fathom = 6 feet
1 cable length = 100 or 120 fathoms
1 nautical mile = 1.15 statute miles
1 league = 3 nautical miles
1 knot = 1 nautical mile per hour

ASTRONOMICAL DISTANCES

1 astronomical unit (AU) = 149,597,870 kilometers = 92,955,800 miles (the mean distance from the earth to the sun)
1 parsec = 3.26 light-years = 3.086×10^{13} kilometers = 1.917×10^{13} miles = 206,265 AU
1 kiloparsec = 1000 parsecs

1 megaparsec = 1,000,000 parsecs
1 light-year = 5.8785 × 10^{12} miles = 9.46053 × 10^{12} kilometers
= 63,240 AU

ANGLE MEASURES

1 minute = 60 seconds
1 degree = 60 minutes = ⅟₃₆₀ of a full revolution
1 sextant = 60 degrees = ⅙ of a full revolution
1 quadrant = 90 degrees = ¼ revolution
1 half-revolution = 180 degrees
1 full revolution = 360 degrees

1 radian = $\dfrac{180}{\pi}$ degrees = 57.29578 degrees

1 degree = $\dfrac{\pi}{180}$ radians = 0.0174533 radians = 1.111 grades

1 grade = 0.9 degrees
2π radians = 1 full revolution = 360 degrees
π radians = half-revolution = 180 degrees
$\dfrac{\pi}{2}$ radians = quarter-revolution = 90 degrees

$\dfrac{\pi}{4}$ radians = 45 degrees

100 grades = 90 degrees
400 grades = 1 full revolution = 360 degrees

KITCHEN MEASURES

1 drop = .065 milliliters
1 teaspoon = 76 drops = 4.93 milliliters = ⅓ tablespoon
= ⅙ fluid ounce
1 tablespoon = 3 teaspoons = ½ fluid ounce = 14.79 milliliters
1 cup = 16 tablespoons = 8 fluid ounces = ½ pint = ¼ quart
= 0.24 liters
1 pint = 2 cups = 16 fluid ounces = ½ quart = 0.473 liter
1 quart = 2 pints = 4 cups = ¼ gallon = 0.945 liter
1 gallon = 4 quarts = 16 cups = 8 pints = 3.785 liters
1 peck = 8 (dry) quarts
1 bushel = 4 pecks

PAPER MEASURES

1 quire = 24 sheets
1 ream = 24 quires = 480 sheets
1 commercial ream = 500 sheets

MISCELLANEOUS AMOUNTS

a tithe = $\frac{1}{10}$ part
a pair = 2 units
a dozen = 12 units
a baker's dozen = 13 units
a score = 20 units
a gross = 144 units = 12 dozen
a myriad = 10,000 units

The Metric System

The basic metric units of length, mass, time, and temperature are the *meter,* the *kilogram,* the *second,* and the *kelvin.* All the other units are derived from these and are called *secondary units.* A liter, for example, is the volume of 1 kilogram of water, and thus is equivalent to 1000 cubic centimeters. Divisions of metric units are formed by adding Greek prefixes for decimal multiples and Latin prefixes for decimal subdivisions.

BASE UNITS OF MEASUREMENT

Type of measurement	*Unit*
Length	meter (m)
Volume	liter (l)
Mass	kilogram (kg)
Time	second (s)
Force	newton (N)
Work	joule (J)
Power	watt (W)
Temperature	kelvin (K)
Electric current	ampere (A)
Light intensity	candela (cd)
Quantity (chemical)	mole (mol)
Frequency	hertz (Hz)

METRIC PREFIXES

Prefix	Meaning
atto-(a)	quintillionth part ($\times 10^{-18}$)
femto-(f)	quadrillionth part ($\times 10^{-15}$)
pico-(p)	trillionth part ($\times 10^{-12}$)
nano-(n)	billionth of ($\times 10^{-9}$)
micro-(μ)	millionth of ($\times 10^{-6}$)
milli-(m)	thousandth of ($\times 10^{-3}$)
centi-(c)	hundredth of ($\times 10^{-2}$)
deci-(d)	tenth of ($\times 10^{-1}$)
deka-/deca-(da)	ten times ($\times 10^{1}$)
hecto-(h)	hundred times ($\times 10^{2}$)
kilo-(k)	thousand times ($\times 10^{3}$)
mega-(M)	million times ($\times 10^{6}$)
giga-(G)	billion times ($\times 10^{9}$)
tera-(T)	trillion times ($\times 10^{12}$)
peta-(P)	quadrillion times ($\times 10^{15}$)
exa-(E)	quintillion times ($\times 10^{18}$)

LINEAR MEASURES

	No. of meters	Approx. U.S. equivalent
kilometer (km)	1000	0.62 mile
hectometer (hm)	100	328.08 feet
decameter (dam)	10	32.81 feet
meter (m)		3.28 feet or 39.37 inches
decimeter (dm)	0.1	3.94 inches
centimeter (cm)	0.01	0.39 inch
millimeter (mm)	0.001	0.039 inch
micrometer (μm)	0.000001	0.000039 inch

AREA MEASURES

	No. of square meters	U.S. equivalent
square kilometer (km^2)	1,000,000	0.3861 square mile
hectare (ha)	10,000	2.47 acres
are (a)	100	119.6 square yards
square meter (m^2)		1.196 square yards
square centimeter (cm^2)	0.0001	0.155 square inch

CAPACITY

	No. of liters	U.S. equivalent
kiloliter (kl)	1000	264 gallons
hectoliter (hl)	100	26.4 gallons
decaliter (dkl)	10	2.64 gallons
liter (l)		1.057 quarts
deciliter (dl)	0.1	3.4 fluid ounces
centiliter (cl)	0.01	0.34 fluid ounce
milliliter (ml)	0.001	0.034 fluid ounce
microliter (μl)	0.000001	0.000034 fluid ounce

MASS AND WEIGHT

	No. of grams	U.S. equivalent
metric ton (t)	1,000,000	1.102 short tons
kilogram (kg)	1000	2.2046 pounds
hectogram (hg)	100	3.527 ounces
decagram (dkg)	10	0.353 ounce
gram (g)		0.035 ounce
decigram (dg)	0.1	0.0035 ounce
centigram (cg)	0.01	0.00035 ounce
milligram (mg)	0.001	0.000035 ounce
microgram (μg)	0.000001	0.000000035 ounce

Conversion Tables

LENGTH

To convert	multiply by
inches to millimeters	25.4
inches to centimeters	2.54
feet to meters	0.3048
yards to meters	0.9144
miles to kilometers	1.61
millimeters to inches	.03937
centimeters to inches	0.3937
meters to feet	3.28
meters to yards	1.0936
kilometers to miles	0.621

WEIGHT

To convert	multiply by
ounces to grams	28.35
pounds to kilograms	0.4536
grams to ounces	0.03527
kilograms to pounds	2.2046

AREA

To convert	multiply by
square inches to square centimeters	6.4516
square feet to square meters	0.0929
square yards to square meters	0.836
square miles to square kilometers	2.5887
acres to hectares (square hectometers)	0.4047
square centimeters to square inches	0.155
square meters to square feet	10.764
square meters to square yards	1.196
square kilometers to square miles	0.3863
hectares (square hectometers) to acres	2.471

VOLUME

To convert	multiply by
cubic inches to cubic centimeters	16.387
cubic inches to liters	0.016387
cubic feet to cubic meters	0.0283
cubic yards to cubic meters	0.7646
cubic centimeters to cubic inches	0.061
liters to cubic inches	61.024
cubic meters to cubic feet	35.3357
cubic meters to cubic yards	1.3079

CAPACITY

To convert	multiply by
teaspoons to milliliters	4.93
tablespoons to milliliters	14.97
fluid ounces to milliliters	29.57
cups to liters	0.24
pints to liters	0.47

quarts to liters	0.95
gallons to liters	3.78
gallons to cubic meters	0.003785
milliliters to teaspoons	0.203
milliliters to tablespoons	0.067
milliliters to fluid ounces	0.034
liters to cups	4.17
liters to pints	2.13
liters to quarts	1.0565
liters to gallons	0.26412
cubic meters to gallons	264.172

MASS AND WEIGHT

To convert	multiply by
ounces to grams	28.3495
pounds to kilograms	0.4536
short tons (2000 lbs.) to metric tons	0.907
grams to ounces	0.03527
kilograms to pounds	2.205
metric tons to short tons	1.1023

TEMPERATURE

To convert	
Fahrenheit to Celsius	subtract 32, then multiply by $5/9$
Celsius to Fahrenheit	multiply by $9/5$, then add 32
Fahrenheit to Rankine	add 459.67
Rankine to Fahrenheit	subtract 459.67
Celsius to Kelvin	add 273.15
Kelvin to Celsius	subtract 273.15
Celsius to Réaumur	multiply by $4/5$
Réaumur to Celsius	multiply by $5/4$

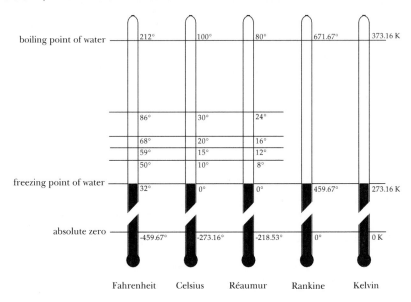

	Fahrenheit	Celsius	Réaumur	Rankine	Kelvin
boiling point of water	212°	100°	80°	671.67°	373.16 K
	86°	30°	24°		
	68°	20°	16°		
	59°	15°	12°		
	50°	10°	8°		
freezing point of water	32°	0°	0°	459.67°	273.16 K
absolute zero	-459.67°	-273.16°	-218.53°	0°	0 K

ENERGY

To convert	multiply by
horsepower to kilowatts	0.7457
calories to joules	4.187
kilocalories to kilojoules	4.187
kilowatts to horsepower	1.341
joules to calories	0.239
kilojoules to kilocalories	0.239

GAS MILEAGE

To convert	multiply by
miles per gallon to kilometers per liter	0.425
kilometers per liter to miles per gallon	2.353

VELOCITY

To convert	multiply by
miles per hour to kilometers per hour	1.61
feet per second to miles per hour	$^{15}/_{22}$
knots to miles per hour	1.151
knots to kilometers per hour	1.85

kilometers per hour to miles per hour	0.621
miles per hour to feet per second	$^{22}/_{15}$
miles per hour to knots	0.869
kilometers per hour to knots	0.540

KITCHEN CONVERSIONS

To convert	multiply by
ounces to grams	28.35
pounds to kilograms	0.45
teaspoons to milliliters	5
tablespoons to milliliters	15
fluid ounces to milliliters	30
cups to liters	0.24
pints to liters	0.47
quarts to liters	0.95
gallons to liters	3.8
grams to ounces	0.035
kilograms to pounds	2.2
milliliters to teaspoons	0.2
milliliters to tablespoons	0.067
milliliters to fluid ounces	0.033
liters to cups	4.17
liters to pints	2.13
liters to quarts	1.05
liters to gallons	0.26

Schedules for Installment Loans ———————

MORTGAGE PAYMENTS

This table gives monthly payments for a $100,000 mortgage. To calculate the monthly payment for any mortgage, divide the amount borrowed by 100,000 and multiply by the number in the table that corresponds to the desired rate and term. (For a 30-year $125,000 mortgage at 8.5% per year: *monthly payment* = $(^{125,000}/_{100,000}) \times 768.91 = \$961.14$.)

Term of Mortgage (in years)

Rate	5	10	15	20	25	30
5.0	1887.09	1060.64	790.78	659.95	584.58	536.82
5.5	1910.12	1085.26	817.08	687.89	614.09	567.79
6.0	1933.28	1110.21	843.86	716.43	644.30	599.55
6.5	1956.61	1135.48	871.11	745.57	675.21	632.07
7.0	1980.12	1161.08	898.83	775.30	706.78	665.30
7.5	2003.79	1187.02	927.01	805.59	738.99	699.21
8.0	2027.64	1213.28	955.65	836.44	771.82	733.76
8.5	2051.65	1239.86	984.74	867.82	805.23	768.91
9.0	2075.84	1266.76	1014.27	899.73	839.20	804.85
9.5	2100.19	1293.98	1044.22	932.13	873.70	840.85
10.0	2124.70	1321.51	1074.61	965.02	908.70	877.57
10.5	2149.39	1349.35	1105.40	998.38	944.18	914.74
11.0	2174.24	1377.50	1136.60	1032.19	980.11	952.32
11.5	2199.26	1405.95	1198.19	1066.43	1016.47	990.29
12.0	2224.44	1434.71	1200.17	1101.09	1053.22	1028.61
12.5	2249.79	1463.76	1232.52	1136.14	1090.35	1067.26
13.0	2275.31	1493.11	1265.24	1171.58	1127.84	1106.20
13.5	2300.98	1522.74	1298.32	1207.37	1165.64	1145.41
14.0	2326.83	1552.66	1331.74	1243.52	1203.76	1184.87
14.5	2358.83	1582.87	1365.50	1280.00	1242.16	1224.56
15.0	2378.99	1613.35	1399.59	1316.79	1280.83	1264.44

MONTHLY PAYMENTS ON SHORT-TERM LOANS

This table gives monthly payments for a loan of $1000, with terms from 2 to 5 years. The total interest figure is the amount of interest paid over the life of the loan. To find the monthly payment *(pmt)* and total interest paid *(int)* for any loan, divide the loan amount by 1000 and multiply by the appropriate figures from the table. (For a 5-year car loan of $12,000 with an 8% annual percentage rate: *monthly payment* = ($12,000/1,000) × 20.28 = $243.36; *total interest paid* = ($12,000/1,000) × 216.80 = $2,601.60; *principal + interest* = $12,000 + $2,601.60 = $14,601.60. The buyer will ultimately pay $14,601.60 for the loan.)

Annual rate	2 yrs Total pmt	int	3 yrs Total pmt	int	4 yrs Total pmt	int	5 yrs Total pmt	int
7%	44.77	74.48	30.88	111.68	23.95	149.60	19.80	188.00
8%	45.23	85.52	31.34	128.24	24.41	171.68	20.28	216.80
9%	45.69	96.56	31.80	144.80	24.89	194.72	20.76	245.60
10%	46.15	107.60	32.27	161.72	25.36	217.28	21.24	274.40
11%	46.61	118.64	32.74	178.64	25.85	240.80	21.74	304.40
12%	47.07	129.64	33.21	195.56	26.33	263.84	22.24	334.40
13%	47.54	140.96	33.69	212.84	26.83	287.84	22.75	365.00
14%	48.01	152.24	34.18	230.48	27.33	311.84	23.27	396.20
15%	48.49	163.76	34.67	248.12	27.83	335.84	23.79	427.40
16%	48.96	175.04	35.16	265.76	28.34	360.32	24.31	458.60
17%	49.44	186.56	35.65	283.40	28.86	385.28	24.85	491.00
18%	49.92	198.08	36.15	301.40	29.38	410.24	25.39	523.40
19%	50.41	209.84	36.66	319.76	29.90	435.20	25.94	556.40
20%	50.90	221.60	37.16	337.76	30.34	460.64	26.49	589.40

Physical Constants

Velocity of light in a vacuum = 1.08 billion kilometers per hour or 670 million miles per hour

Velocity of sound at sea level at 0°C (called Mach 1) = 331.5 meters per second = 1087 feet per second = 741 miles per hour

Velocity of sound at sea level at 60°C= 340.3 meters per second = 1116 feet per second = 761 miles per hour

Velocity of sound through water at 60°C = 1450 meters per second = 4750 feet per second = 3240 miles per hour

Equatorial radius of the earth = 3963.34 statute miles = 6378.338 kilometers

Circumference at the equator = 24,900 miles

The golden ratio = 1.61803 to 1

Fraction–Decimal–Percent Conversions

Fraction to decimal: Divide the numerator by the denominator.

Decimal to fraction: Write the digits to the right of the decimal point over the power of 10 corresponding to the place value of the last digit.

Decimal to percent: Move the decimal point two places to the right.

Percent to decimal: Move the decimal point two places to the left.

Percent to fraction: Write the percent figure over 100 and reduce.

Fraction to percent: Convert to decimal form, then to percent form. Or, if possible, convert to an equivalent fraction with a denominator of 100; the numerator will be the percent.

Mixed number to improper fraction: Multiply the denominator times the whole number and add the numerator to derive the new numerator, keeping the same denominator.

Improper fraction to mixed number: Divide the denominator into the numerator and express the remainder as a fraction.

Length, Area, and Volume Formulas

Circle

circumference = $2\pi r$

area = πr^2

area of a sector = $\left(\dfrac{\theta}{360} \right) \times \pi r^2$ (degree measure)

$\qquad\qquad$ or $\dfrac{\theta r^2}{2}$ (radian measure)

length of an arc = $\dfrac{\pi r \theta}{180}$ (degree measure)

$\qquad\qquad$ or $r\theta$ (radian measure)

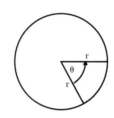

Triangle

area = $\dfrac{1}{2}bh$

Square

area = x^2

perimeter = $4x$

Rectangle

area = (length) × (width) = xy

perimeter = $2x + 2y$

parallelogram

area = $b \times h$

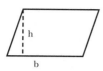

Trapezoid

area = $\frac{1}{2}(a + b) \times h$

Rectangular box

volume = (length) × (width) × (height) = xyz

surface area = $2xy + 2xz + 2yz$

Cylinder

volume = (area of base) × height = $\pi r^2 h$

surface area = $2\pi rh + 2\pi r^2$

Cone

volume = $\frac{1}{3}\pi r^2 h$

surface area (not including base) = πrs

Pyramid (base may be any shape)

volume = $\frac{1}{3}$ (area of base) × height

Sphere

volume = $\frac{4}{3}\pi r^3$

surface area = $4\pi r^2$

Trigonometric Functions and Right Triangles ———

RIGHT-TRIANGLE TRIGONOMETRY

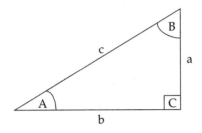

sine of A (sin$A$) = a/c
cosine of A (cos$A$) = b/c
tangent of A (tan$A$) = a/b
cosecant of A (csc$A$) = $1/\text{sin}A$ = c/a
secant of A (sec$A$) = $1/\text{cos}A$ = c/b
cotangent of A (cot$A$) = $1/\text{tan}A$ = b/a
Pythagorean Theorem: $a^2 + b^2 = c^2$

Law of Sines: $\dfrac{a}{\sin A} = \dfrac{b}{\sin B} = \dfrac{c}{\sin C}$

Law of Cosines: $a^2 = b^2 + c^2 - 2 \times b \times c \times \cos A$
$b^2 = a^2 + c^2 - 2 \times a \times c \times \cos B$
$c^2 = a^2 + b^2 - 2 \times a \times b \times \cos C$

Trigonometric Identities

$\sin^2\theta + \cos^2\theta = 1$
$1 + \tan^2\theta = \sec^2\theta$
$1 + \cot^2\theta = \csc^2\theta$

$\sin\theta = \cos(90° - \theta)$
$\cos\theta = \sin(90° - \theta)$
$\tan\theta = \cot(90° - \theta)$
$\cot\theta = \tan(90° - \theta)$

Trigonometric Function Values for Acute Angles

θ (in degrees)	sinθ	cosθ	tanθ	cotθ
0	.0000	1.000	0.000	∞
1	.0175	.9998	.0175	57.2900
2	.0349	.9994	.0349	28.6363
3	.0523	.9986	.0524	19.0811
4	.0698	.9976	.0699	14.3007
5	.0872	.9962	.0875	11.4301
6	.1045	.9945	.1051	9.5144
7	.1219	.9925	.1228	8.1443
8	.1392	.9903	.1405	7.1154
9	.1564	.9877	.1584	6.3138
10	.1736	.9848	.1763	5.6713
11	.1908	.9816	.1944	5.1446
12	.2079	.9781	.2126	4.7046
13	.2250	.9744	.2309	4.3315
14	.2419	.9703	.2493	4.0108
15	.2588	.9659	.2679	3.7321
16	.2756	.9613	.2867	3.4874
17	.2924	.9563	.3057	3.2709
18	.3090	.9511	.3249	3.0777
19	.3256	.9455	.3443	2.9042
20	.3420	.9397	.3640	2.7475
21	.3584	.9336	.3839	2.6051
22	.3746	.9272	.4040	2.4751
23	.3907	.9205	.4245	2.3559
24	.4067	.9135	.4452	2.2460
25	.4226	.9063	.4663	2.1445
26	.4384	.8988	.4877	2.0503
27	.4540	.8910	.5095	1.9626
28	.4695	.8829	.5317	1.8807
29	.4848	.8746	.5543	1.8040
30	.5000	.8660	.5774	1.7321
31	.5150	.8572	.6009	1.6643
32	.5299	.8480	.6249	1.6003
33	.5446	.8387	.6494	1.5399
34	.5592	.8290	.6745	1.4826
35	.5736	.8192	.7002	1.4281
36	.5878	.8090	.7265	1.3764
37	.6018	.7986	.7536	1.3270
38	.6157	.7880	.7813	1.2799

Trigonometric Function Values for Acute Angles—cont'd

θ (in degrees)	sinθ	cosθ	tanθ	cotθ
39	.6293	.7771	.8098	1.2349
40	.6428	.7660	.8391	1.1918
41	.6561	.7547	.8693	1.1504
42	.6691	.7431	.9004	1.1106
43	.6820	.7314	.9325	1.0724
44	.6947	.7193	.9657	1.0355
45	.7071	.7071	1.000	1.0000
	cosθ	sinθ	cotθ	tanθ

Rules for Exponents and Radicals

	example
$a^0 = 1$	$10^0 = 1$
$a^1 = a$	$10^1 = 10$
$a^n = a \times a \times a \times a \times a \ldots \times a$ (n times)	$2^5 = 2 \times 2 \times 2 \times 2 \times 2 = 32$
$a^{-1} = \dfrac{1}{a}$	$2^{-1} = \dfrac{1}{2}$
$a^{-n} = \dfrac{1}{a^n}$	$2^{-5} = \dfrac{1}{2^5} = \dfrac{1}{32}$
$a^{1/n} = \sqrt[n]{a}$	$8^{1/3} = \sqrt[3]{8} = 2$
$a^m \times a^n = a^{m+n}$	$3^2 \times 3^4 = 3^{2+4} = 3^6 = 729$
$a^m \div a^n = a^{m-n}$	$5^4 \div 5^3 = 5^{4-3} = 5^1 = 5$
$(ab)^n = a^n \cdot b^n$	$(5x)^2 = 5^2 \cdot x^2 = 25x^2$
$\sqrt{ab} = \sqrt{a} \cdot \sqrt{b}$	$\sqrt{49x^2} = \sqrt{49} \cdot \sqrt{x^2} = 7x$
$\sqrt{\dfrac{a}{b}} = \dfrac{\sqrt{a}}{\sqrt{b}}$	$\sqrt{\dfrac{64}{81}} = \dfrac{\sqrt{64}}{\sqrt{81}} = \dfrac{8}{9}$

Rules of Thumb for Measuring and Calculating

A dollar bill is 6⅛ inches long and 2⅝ inches wide.

A standard business card is 3½ inches long and 2 inches wide.

A standard credit card is 3⅜ inches long and 2⅛ inches wide.

A dime has a diameter of 9/16 inch, which is also the thickness of a stack of 100 sheets of ordinary paper.

A penny has a diameter of ¾ inch and weighs about 2.5 grams.

A nickel weighs about 5 grams.

A quarter has a diameter of 15/16 inch.

The wire of an ordinary paper clip has a diameter of about 1 millimeter.

One gram is a little more than the weight of an ordinary paper clip.

One centimeter is about the width of an ordinary paper clip.

One liter of water weighs 1 kilogram.

One milliliter of water weighs 1 gram.

A pint of water weighs a little over a pound.

A gallon of water weighs about 8.5 pounds.

A gallon of gasoline weighs about 5.8 pounds.

Divisibility Tests

SIMPLE TESTS

A number is divisible by:	*if:*
2, 4, or 8	if its last digit, last two digits, or last three digits form a number divisible by 2, 4, or 8.
3 or 9	if the sum of the digits is divisible by 3 or 9.
5	if the last digit is 0 or 5.
10	if the last digit is 0.
25	if the last two digits are 00, 25, 50, or 75
11	if the sum of the odd digits (1st, 3rd, etc.) minus the sum of the even digits (2nd, 4th, etc.) is 0 or 11.

OTHER TESTS

7, 11, and 13   Split the number into groups of three (triads) from right to left. Then, from left to right, subtract the first triad (which may have less than three digits) from the second, subtract the result of this from the third, and keep subtracting up to the last triad. If the resulting number is divisible by 7, 11, or 13, then so is the original number. (To test 1,377,558 for divisibility by 7, 11, or 13, subtract 1 from 377 to get 376. Then subtract 376 from 558 to get 182, which is divisible by 7 and 13. Thus the original number is divisible by 7 and 13.)

7, 11, 13, 17, and 19

7: Split off the last digit of the number, double it, and subtract this from the number that remains. Repeat this if necessary until the remainder is recognizable as a multiple of 7. (To test 35,672 for divisibility by 7: Split off the 2, double it to get 4, then subtract this from 3567 to get 3563. Double 3 to get 6, and subtract this from 356 to get 350. Double and subtract the 0 to get 35, which is divisible by 7.)

11: Split off the last digit and subtract it (without doubling) from the remaining digits. Repeat until the result is 0 or 11.

13: Split off the last digit and quadruple it, then *add* this to the remaining number. Repeat until the result is recognizable as a multiple of 13.

17: Split off the last digit, multiply it by 5, and *subtract* this from the remaining number. Repeat until the result is recognizable as a multiple of 17.

19: Split off the last digit, double it, and *add* the result to the remaining number. Repeat until the result is recognizable as a multiple of 19. (To test 67,393 for divisibility by 19, split off the last 3, double it to get 6, then add this to 6739 to get 6745. Double the 5 to get 10, and add this to 674 to get 684. Double the 4 to get 8, and add this to 68 to get 76. Finally, double the 6 to get 12, and add this to 7 to get 19.)

Index